# Pepel & People

# PEPEL & PEOPLE

L. V. Gray

ILLUSTRATED BY RICHARD ORR

LONDON
MICHAEL JOSEPH

First published in Great Britain by
MICHAEL JOSEPH LTD
52 Bedford Square
London, W.C.1
1970



7181 0761 6

Printed in Great Britain by Hollen Street Press
and bound by James Burn at Esher, Surrey

# 'The Day I Shall Never Forget'

It happened on June 18 1968 . . . a day I shall remember for the rest of my life. The day Pepel died, then lived again.

She had gone out for a romp on deck, on the oil tanker *Arduity*. Pepel did this most evenings after tea, while I watched her play hide and seek among the pipes and valves or run along the catwalk to visit the Captain, for whom she had much affection.

The vessel had been carrying spirit and the tank lids had been left open to allow the gas to escape, rendering the tanks gas-free for our next cargo waiting for us at Coryton, in the Thames. I did not know this at the time or I would have been more concerned for the safety of my pet monkey, who was courting death as she played on the tank tops.

Pepel suddenly disappeared from view, but knowing her little tricks of hiding from me, then jumping at me from behind, I did not worry until twenty minutes had elapsed. Captain Reg Watson, who had come on deck to watch Pepel play, was more concerned than I was when he could not see her about. He searched the cabins midships and the forecastle, then looked to see if she had gone on the bridge, where she spent her time during the day. By this time I had also become very worried, thinking that she might have slipped off a rail and gone into the sea.

I was searching the lifeboats aft when I heard the Captain call out that Pepel was down the tanks. I rushed to him on the oil deck as he climbed down to get her thoughts racing through my mind that she was jumping about in the oil after having her nightly shower, but as I looked down the tank I could see her sprawled on a grating at the bottom of the ladder. When Captain Watson reached her I asked him if she was all right. He replied, in an accusing voice, 'She is dead, Tiny' (Tiny is my nickname). I said, 'Oh no, she can't be, she must not be dead!' I leaned forward to take Pepel as he lifted her up to me and knew as I looked at her that he must be right, but I refused to believe it. I was trembling all over as I cuddled her limp little body to mine. Her limbs were rigid, her little stomach was bloated like a balloon with the gas she had inhaled. She had no pulse and she was definitely not breathing. I looked into her glazed, unseeing eyes, tears streaming from my own as I cradled her lifeless body in my arms and said, 'Why did you have to leave me, Pepel?'

After a few moments of panic and despair it suddenly occurred to me that I might bring her back by giving her the 'kiss of life'. I had never done this before and did not really know how to do it, but in my desperation I put my mouth over her foam-flecked lips and blew air into her lungs. The Captain stood by helpless and wishing that he could help me in some way, knowing how much I loved my monkey, just as he had come to be very fond of her. We still stood on the oil deck and half-an-hour must have passed when, just as I was about to give up, Pepel gave a great gasp, then started to breathe very weakly. I said, 'Reg, Reg, she is breathing!' He said, 'Quick, get her to your cabin and wrap her in a blanket'. I rushed to do as he said, still breathing into her mouth. After a while her eyes flickered and became unglazed and she began to clench her little fists as if she was trying her best to help me. I could feel her heart start to beat again and her stomach became warm and less bloated as the gas came out of her. I covered her up on my bunk when she started to breathe regularly, exactly one hour after she was brought out of the tank.

Her eyes became clear and she looked around, yawning, as if to say, What is all the fuss about? My happiness was complete when Captain Watson came into the cabin and Pepel chattered to him as she usually did, not realising that he had just helped to save her life. My tears turned to laughter when she tried to get up to go to the Captain, only to fall out of the bunk. She was quite drunk and remained so for several hours until the gas had left her system.

I sat up with her until three o'clock the next morning, trying to coax her into drinking something. She finally sipped a drop of lime juice and even nibbled at a sweet biscuit. She had been visited throughout the night by members of the ship's company, all of whom asked how 'Our Pepel' was getting on.

There was no sleep for me for many a night after. I lay on my bunk listening to the Captain saying, 'She is dead, Tiny. She is dead, Tiny'. I shall never be able to repay him enough for what he did for my little friend. I shall always remember the moment of panic after he had brought her out of the tank. Every little incident that happened during the four years Pepel had been my constant companion flashed through my mind. All the happy moments we have had together and the comfort she has afforded me became real and I knew that if she could not be revived there would be a great void in my life, an emptiness I could never get used to. The deep and trusting love she has for me could never be replaced, by human or by animal.

Gratitude prompted me to write the following story about Pepel. Talking about her has never been difficult; there is always plenty to talk about where Pepel is concerned. Writing about her has been even more of a pleasure to me because I know that her story, the story of a very beautiful and unusual monkey, will be of interest to animal lovers all over the world, especially to those who have had the pleasure of meeting Pepel and those who are fortunate in having a pet monkey of their own.

The story of Pepel may also be of some interest to the many scientists who daily experiment with monkeys less fortunate than Pepel. Their task of finding out how monkeys tick must surely be a gruesome one, to say the least. Last year, 'Experiments on living animals numbered 4,755,680, says a report published today by the Stationery Office. Eighty-seven per cent of the experiments were carried out without anaesthetic'. (Quoted from *The Sun*, August 14, 1968.)

What a tragedy it is for these unfortunate creatures when they are taken from the sunshine of their homeland, to end up on the cold slab of the vivisector. Beautiful, super-intelligent creatures, always bubbling with life, who are cut up and thrown into a trash bin, all for the sake of the human beings they unfortunately resemble!

I realise that certain experiments must be carried out and that nothing we ordinary people might say or do would have any effect towards stopping such experiments. However, I hope that the story of Pepel will have some effect on the conscience of any person engaged in such practices, and that her brothers and sisters will be treated with some compassion.

Pepel has so far lived a cosmopolitan life and has met all kinds of people in many different parts of the world. As it has been impossible to write about her life among people without including them in the story, deciding on a title for the book was easy, it is simply called 'PEPEL AND PEOPLE'.

L. V. GRAY (TINY)

# Chapter One

Of all the trips I made to Africa during my life at sea the most memorable was the voyage to Sierra Leone on the m.v. *River Afton* in March 1964. The passenger liners I had sailed in previously used to call at the more romantic sounding places, like Cape Town, Port Elizabeth, Durban, Lorenço Marques and Beira, all very beautiful South and East African ports. Especially Durban, with its rich, breath-taking jungle coastline and crystal blue waters sweeping into what must surely be one of the world's most beautiful harbours until it reaches the quayside, where the famous Zulu warriors are waiting with their rickshaws to show you the delights of a city that never really sleeps.

I joined the *River Afton* when she berthed at Ijmuiden in Holland, loaded with a cargo of iron ore. It was drizzling with rain at the time and everything was muddy and black, as black as the skins of the crew who were handling the ropes and wires as the vessel came alongside the huge electric cranes, waiting to discharge her equally black bowels. The first grab slammed into the ore even as I walked up the gangway and I thought, 'They must want an extra quick turn-round this trip'.

However, the ship was not due to leave Holland for a couple of weeks, as she had to go to Rotterdam for a refit before she sailed once again to West Africa. This gave me time to settle in to my job as chief steward and to get to know the crew, who were all natives of Sierra Leone. Sambo, my head steward and 'Captain's Tiger', introduced me to the catering crew. I could see, as they looked at my nineteen stone hulk, that they were wondering what kind of a boss I would be. When I told them that there would be extra overtime paid if they gave the ship a good clean up they were all smiles. From that day on I had the best African crew at sea.

The chief cook and I became firm friends. He used to get me to write his letters for him on the 'machine', my typewriter. He was always promising to make for me a 'Palm Chop', a native dish made with palm oil and clams, among other ingredients, and although I believe it to be a very tasty dish, I never did manage to get it from him. I shall always remember Abdullah for his ever-smiling face and the huge quantities of rice he used to eat, smothered with a mixture of chicken fat and tomato puree. He took great pride in telling everyone that the chief steward was a good man because, apart from the weekly inspection, I left him and his staff alone to get on with the job. Being a chef myself, I have always believed in too many cooks spoiling the broth.

We eventually sailed for Freetown on March 15, 1964. As we left the snow and frost of the Netherlands behind us I had a feeling of gratitude towards the shipowner who had made it possible for me to go to a warmer, if not healthier, climate. The crew, who had spent so many miserable weeks in a severe cold climate, were also very glad to be sailing into the sunshine and going home. During the long voyage across the Atlantic it did not take me long to understand these home-loving Africans who, in their spare time, were always busy carefully wrapping the gifts they had bought in Holland for their families. They would come to me with all their problems and their aches and pains. I listened with sympathy to Sonny, the galley boy, who told of the rats which killed his chickens at home in Freetown, and to the second cook who was due for a beating from his father because he had failed to get him a rocking chair from Holland. I cured all aches and pains with that famous pill the Aspro. It was much better, they said, than 'the horrible black-jack that the other Chief gave to us'. Through treating these good-natured Sierra-Leoneans like men and with a mixture of discipline and kindness, I became their friend and I was soon to be known as the 'White Kaffir'. They came to have so much faith in me that it became rather embarrassing at times. If one of the other officers told them off for some reason, they used to come to me for support. They seemed to recognise the fact that I felt humble in their presence, knowing that their blood was just as red as my own. A couple of trips later I was invited to their village . . . a place called 'Pepel'.

# Chapter Two

The sun was peeping at us on the horizon as the *River Afton* arrived at Freetown, where the African crew are allowed to go home while a relief crew come aboard to take the vessel up the river to Pepel, the loading port for the iron ore which is transported by trains from the mines further inland. The arrival of the 'Baby Afton', as she was called locally, was an event I shall never forget. As the anchor plunged into the Bay the decks were aswarm with happy and excited black men, laughing and jostling each other while they waited for the relief men to come out on the barge they all knew so well. I could hardly believe my eyes when I went on deck to meet the Customs officials, who were usually the first to come on board, for there, stacked all along the starboard side of the ship, was the biggest collection of furniture and junk I have ever seen: sideboards, chairs, old-fashioned bedsteads, mattresses, dining tables, bicycles, radio sets and bric-a-brac of every description lay all over the deck. The ship looked like a floating junk shop. The chief engineer told me that it was the same every trip and that the crew scoured the secondhand shops of Rotterdam for this rubbish and actually paid duty on it in Freetown. I could hardly believe that these items could be of any value, except perhaps on a bonfire, but the chief engineer assured me that these shrewd Sierra Leoneans made a fortune out of the junk business.

With the pilots on board and after the sea-going crew had gone ashore with the rest of the officials, the ship made her way up river, leaving the clean air of the bay to make very slow progress into the humid heat of the jungle for her cargo of ore. Being a large person I felt the heat more than most on board. I was certainly grateful for having a private shower to cool off in.

The usual 'bum-boat' men were waiting for us as the anchor was dropped at Pepel, offering fruit, muskrat skins, parrots, monkeys, and all kinds of local souvenirs, in exchange for cigarettes or any cast-off clothes we had to part with. All transactions were made under the beady eye of the Customs official who had travelled from Freetown with us. He did not mind what went ashore as long as he was suitably rewarded for looking the other way. Being chief steward I was looked upon with great respect by these men, who usually had a sad story to tell, knowing that I had a soft heart and the keys to the stores. I did not know at the time but I was soon to be very thankful that I had kept on the right side of this particular Customs man.

On my first trip to Pepel I had made friends with the berthing master and his staff. The berthing master was a Scotsman with a love for kippers, a commodity apparently not obtainable in that part of the world unless it could be scrounged from the ships calling there. Anyway, my friend had asked me on the previous trip to take him a few kippers back for himself and a few other kipper-hungry Scots at Pepel. I took a whole box for them and it turned out that those kippers became my passport to enable me to go ashore.

It seems that the colonials were fed up with having their club smashed up by visiting seamen and as the same seamen had caused trouble with the women in the local village, all Europeans had been stopped from going ashore at Pepel. This applied to officers and crew alike. Needless to say, I was very grateful when I was invited ashore by my kipper-loving friends who 'stood bond' for me. The other officers good humouredly pulled my leg about me being the only lucky devil allowed to go ashore, while they had to sweat it out on board swatting the millions of insects that had come to keep us company. The only thing I was worrying about was how to tell them why there would be no kippers for breakfast on the return voyage!

Little did I realise it then, but the invitation to spend an evening ashore in Pepel was to change the course of my life completely. Until then I had been a very lonely bachelor interested only in my work and going to sea. Circumstances had prevented me from marrying earlier in life and I was no longer interested in getting married and having a family. Instead, I had chosen the whole world for my family in the many sincere friends I had made and re-visited from time to time in various countries.

Pepel changed all that. Pepel gave me a responsibility I can never shirk, no matter what may happen in the future. Pepel enriched my life by giving me something to love. The People of Pepel gave me PEPEL.

# Chapter Three

Going ashore at Pepel is not like doing the same thing in Durban. Although the scenery is very beautiful with its rich green foliage and swaying palm trees towering above little shacks built among sand dunes, the only sign of life on the seashore is the occasional woman who can be seen carrying a basket of something or other to and from the European bungalows. At night there are only the twinkling lights to be seen and the only sound to be heard is the long blast of the train whistles above the murmur of jungle as the wagons loaded with ore are shunted to the jetty. There are no rickshaw boys or taxis, no electric signs or nightclubs. In Pepel there is just a great long jetty followed by an equally long dirt track to walk along before you reach the European settlement; then, if you want to go to the African village you have to walk across a paddy field and along another dirt track. Several people have been killed by snakes as they walked along this path to the village. I

B

thought my friends were pulling my leg when they told me that I should carry a torch at night to frighten the snakes away. They told me this when I said that I would like to go and see what an African village looked like. After I had been provided with a torch and had been given the directions to the village, I set out along this so-called treacherous footpath. The only living thing I encountered was a great land crab, but I was thankful for the comfortable feeling the torch gave to me.

After leaving the paddy field the pathway went through about two hundred yards of jungle until it reached a railway embankment. The village lay on the other side of this embankment, but even before I crossed over the track I could hear pop music blaring out from what I thought would be the usual honkytonk bar, not uncommon where seamen are found. I was not quite prepared for what I stumbled into after I had gone down the other side of the embankment.

Dancing to the music being played on what must have been the oldest 'His Master's Voice' record player, complete with the old-fashioned horn, were a group of naked boys and half-naked girls, all twisting and shaking to an Elvis Presley record as I have never seen it done before. I stood there amazed. After my long walk in the tropical night my white shirt and shorts were wet through with sweat as I stood looking at these kids performing acrobats on a dirt floor in front of the bar where the music was coming from. I could not help but envy the youngsters and wished that I was a few stone lighter so that I could join in the fun.

The bar was just a small wooden hut with a corrugated tin roof. It had a small verandah at the front and what seemed to be chicken houses grouped at the rear. I learned later that these buildings were in fact the living quarters, each one containing just a bed, a chair and chest of drawers, etc. The bar counter was just a wide plank of wood placed across two empty Heineken boxes. Two basket chairs stood on the verandah and sitting in one of them was the Customs officer who had travelled up from Freetown with us. Like the kids, who had stopped dancing to watch my arrival, his black skin was dry and he looked as cool as the iced Heineken lager he drank from the bottle. I flopped down beside him and dabbed my sweating bald head with an already sopping wet handkerchief. After 'Mama' had supplied me with a bottle, I too sampled a cold lager and wondered how they kept it so cool, seeing that the only form of light in the bar was an oil lamp. I never did find out because the dancing had started again, after my friend the Customs man had wound the antique gramophone up. Then, while the lads and lassies threw their limbs about, the smaller kiddies started crawling all over me. I made the fatal mistake of giving a few Dutch coins to one of them and in no time at all I had all the kids of the village around me.

Then the mother of Pepel came on the scene. I was swigging my fourth bottle of Heineken when a naked boy of about twelve years of age came out of the darkness. He was the son of the woman who kept the bar and he had, on the end of a length of flex, a very beautiful monkey. I had never seen one so close before. Although I have always been very fond of animals I had never been to a zoo in my life; the only monkeys I had seen were those used by photographers at the seaside. I fell in love with this one at first sight and just as the Customs man was about to tell me not to try to handle her, the monkey jumped on to my shoulder and urinated on me, as if I was not wet enough! With the Customs man acting as my interpreter I asked 'Mama' what kind of monkey it was. She replied that it was a ground monkey but was known locally as a 'Golden

Tailed Mango Monkey'. She asked me if I would like a baby monkey and I replied by saying that I would like to have one very much, as long as it was a small one. She smiled and spoke to her little boy, who disappeared again into the night to fetch what turned out to be the new born baby of her own monkey. Just a little bundle of fur with beautiful brown eyes and only five days old.

The little boy came back with his sister, who was carrying my little gift nestled up between her small bare breasts. The girl was crying because her mother was making her give up the baby monkey. I felt awful about taking it, but later, when I was told by Sambo that the monkey would have been eaten, I was glad to have saved Pepel for a better fate. I gave 'Mama' a carton of Lucky Strike for the little mite, who just covered the palm of my hand. By that time two of the village policemen had arrived for a drink and it was during the polite conversation that ensued, while the Lucky Strikes were being passed around, that I dropped a clanger. One of the policemen asked me what I intended to call the monkey. After a moment's thought I told him that I would call it Sambo, after my head steward. They very nearly put me in the local clink for insulting Sambo by wanting to give the same name to a mere monkey. I tried to explain to them that in England it was a common thing for animals to have the same names as human beings, but I could see that they did not like it. It cost me the price of several bottles of beer before we got back on to friendly terms.

Pepel, as I later decided to call her, was fast asleep as I made my way back to the ship, escorted by the Customs officer and the now slightly drunken policemen. They took me a quicker way back along the railway track, where I saw my first snake slithering away from the torchlights. We said goodbye to the policemen at the end of the jetty and walked aboard the *River Afton*, where everyone was asleep except for the watchman on the gangway.

I made a bed for Pepel in a cardboard box in which I had put a couple of my vests. She was still asleep as I covered her with a large bath towel. I put the box and contents in the bathroom and retired to my bunk, wondering what I would say to the Captain the next day, or what he would say to me when he knew that I had brought a monkey back on board with me. Anyway, for the moment I was too full of Heineken's Dutch courage to worry unduly and feeling like a schoolboy with a new pet I went to sleep worrying about how to feed it.

Pepel was awake before me the next morning, for when my steward brought in the morning tea she was screeching her little lungs out in the bathroom. Munsra, the steward, laughed his head off when I opened the bathroom door to show him the reason for all the noise! Pepel was perched on the shower curtain rail and had a look of defiance on her face as I approached her. As I lifted her down she tried to bite me, which was understandable, as she had never seen a giant like me before. Her attempted bite had no effect because being only a few days old she had no teeth. However, just to show her who meant to be the boss, I gave her a cuff on the bottom and put her back in her box.

Although I tried to keep her presence a secret until we were safely at sea, the word soon got around that there was a monkey on board. The first day on the ship must have been a very disturbing one for Pepel with all kinds of black and white faces coming in to look at her. The Captain, a very nice chap called Chinnery, was a bit dubious at first. It was his first command and I suppose he was thinking about port health restrictions when we got back to Holland, or whether there would be any objections to the monkey being on board from other officers, as I later found to be the case on other vessels. Captain Chinnery allowed me to keep Pepel, however, and later became very fond of her, just as almost everyone on board did. It was a great relief to me because even at that early stage I would have hated to part with my new little friend, who sat in her strange bed, looking at me as if to say, 'Well, have I passed inspection?'

Soon, however, the excitement of having a pet on board gave way to the business of getting the ship back to Freetown, where the regular crew were waiting to come back on board. I was very pleased to see the familiar faces of my catering crew scrambling up the gangway from the barge. After their short respite at home I knew that they would soon get the department back to normal. Sambo had brought me a gift of two potted palms, which I put in the saloon. My own steward brought back my spare white uniforms which he had taken home and laundered. All the African boys crowded round my cabin door to catch a glimpse of the baby monkey as she sat on my armchair. Unfortunately, Pepel did not care much for her friendly audience and hid from view at the back of the chair. Even to this day she does not like people with dark skins and will hide from them, or attempt to bite them if she gets half a chance. Although she has sometimes allowed one or two black people to coax her, she has an inborn fear and hatred of people who come from her own country. Perhaps it is because her instinct tells her that black people hunt and eat monkeys, or may be she remembers the cruel way in which her mother was tethered and dragged along at the end of a length of flex back in Pepel.

# Chapter Four

It was as we were steaming back to Holland that I noticed something was wrong with Pepel's right ear. It was swollen and inflamed and I noticed that she kept rubbing it with her hand. On taking a closer look I could see a puncture, about the size of a pin-head at the top of her ear and the swelling seemed to be moving up and down as if it were breathing. I bathed the ear with a solution of warm water and T.C.P. and while doing so I noticed little bubbles coming out of the puncture. It looked as if something was trying to get out of the hole. After an apology to Pepel I began to probe inside the hole with a pair of pointed tweezers and after several attempts I pulled out the cause of her misery, a horrible grub which I learned later was a tsetse grub, planted in Pepel's ear by a tsetse fly. All through the operation Pepel had never made a murmur and, to show her gratitude, she flung her little arms around my neck and sobbed like a baby. To this day Pepel has a small black scar at the top of her right ear where the grub came out.

Pepel soon made herself at home in my cabin and as my quarters were situated aft, just below the boat deck and next door to the saloon, she had plenty of space to play in where she would be no trouble to anyone. For the first few days she never left the cabin unless it was to follow me, as she has followed me ever since. Soon, however, she ventured a little way beyond the cabin door to the door leading out on deck. She would go no further than that though, as the noise of the engine frightened her so much. I would take her further aft on the same deck where it was quieter and she would sit or play in the sunshine while I sat in a deck-chair to watch her. The only thing that worried me was her habit of sitting on the ship's rail, without a care in the world, as I sat gaping with my heart in my mouth, not daring to speak to her in case she fell into the sea that was rushing by beneath her. Like all monkeys, Pepel has a wonderful sense of balance and can even sit on a clothes line with ease. I have only seen her lose her balance once. She was slightly drunk at the time and toppled over as she attempted to jump from one chair to another.

One of the ship's firemen was also an expert tailor. He made all my tropical shirts and trousers on an old-fashioned sewing machine he kept in his cabin. One day he came to me and asked if he could measure Pepel for a coat he would make for her. Strangely enough she allowed a black man to handle her as he ran the tape over her and in due course he presented me with a beautiful fur coat, saying, 'The little monkey will need this when she is in Holland'. It was a wonderful gesture on his part and I rewarded him with a bottle of rum, for which Pepel later became the owner of a complete wardrobe of little clothes made by this man, who knew that my pet monkey was something special to me.

Doing my paper work became a problem because Pepel

would not leave me alone. She thought she had the right to use my typewriter at the same time as myself and when I tried to do my figures and requisition sheets she was a constant distraction. Also, when my back was turned she delighted in pulling out the typewriter ribbon and would sit with it piled around her while she tied it in knots. The chief officer said that Pepel knew more knots than the crew! She loved to get hold of my pens, especially the ball-pointed ones, and quickly learned how to unscrew the top and withdraw the refill.

When things began to get really out of hand I asked the ship's carpenter to make a cage for me to put Pepel in while I was busy. This eased the situation a little, but I could not stand hearing her sobbing as she cried to be let out. Needless to say, the carpenter went to all his trouble for nothing and Pepel was happy to see me throw her awful prison over the side. Strangely enough it was a souvenir bought from one of the bumboat men at Pepel that finally solved the problem. One of the junior engineers had bought a muskrat skin and while he was showing it to me Pepel hid behind my desk, making frightened clacking noises. She would not go near the engineer who was holding the skin. I took it from him and put it on my desk. Pepel jumped on to a coffee table at the other side of the cabin and stood upright as she chattered away in protest at the lifeless intruder lying next to her typewriter! The engineer agreed to lend me the skin for the rest of the trip and from then onwards I had at least one corner of the cabin to call my own.

People always remark how clean Pepel is and although I take some of the credit for keeping her clean now, it was on her own initiative that she started to bathe regularly. She was still sleeping in the bathroom and every time I took a shower, sometimes three or four times a day in the hot weather, she would look at me curiously as I lathered myself and made the usual spluttering noises we humans make under a shower. Curiosity got the better of her one day when she decided to get a little closer and sit on the faucet to look down at me. She then put a hand gingerly under the spray and let the water trickle through her fingers. She did this a few times before she plucked up enough courage to sit on my shoulder and allow the water to fall over her. It was just a matter of time then for me to coax her into allowing me to lather her with the soap. She got to like the bath so much that sometimes, when I had finished drying her, she jumped back under the shower. Although I became annoyed at this I could not help laughing at her antics and the expression on her face when reprimanded. Now there is nothing Pepel likes better than to swim in the slipper bath and to dive from the side of the bath after the soap.

Pepel started to sleep in my bunk before we reached Holland. It happened one night after a party I gave to some of the officers who were off watch. She had gone to bed in the bathroom earlier in the evening after a romp at the party, but I suppose the continuous use of the toilet spoiled her rest. After my guests had departed for their own bunks, a little drunk after emptying all my bottles, I turned in myself, but as I too was a little under the influence I forgot to close the bathroom door. The next morning I woke up to find Pepel with her head on my pillow and one of her hands on my face. She had snuggled right under the bedclothes just as if she had been used to sleeping in a bed for years. She has slept with me all the time since then. She sleeps right through the night and is not a bit of trouble; in fact, the only movement she makes is when I turn over and she moves from one side of me to the other and settles down in the crook of my knees. I find Pepel better than a hot water bottle in the winter!

# Chapter Five

After only a few hours in Ijmuiden, during which time I was kept busy with storing the ship and entertaining Customs officers, we returned on what was to be my last voyage to Africa on the *River Afton*. It was during this outward-bound journey that Pepel started to have teething trouble. Just like a baby she wanted to be nursed all the time as she cried with the pain of her first teeth pushing through sore gums. I remembered someone telling me about painting children's gums with whisky when teething trouble began, so I did the same to Pepel. I used cotton wool soaked in a glass of Teacher's and dabbed it on her gums. The first application took her breath away and she looked at me as if to say, 'What the hell are you trying to do to me?' After that, however, she actually looked forward to having her gums massaged with whisky and used to retrieve the cotton wool from the wastepaper basket in order to suck the last drop of whisky from it. Her teeth began to show at the front, just like little human teeth and milk white.

I had been feeding Pepel on diluted Nestlés full cream sweetened milk laced with a tot of Navy rum, also rice pudding and any soft fruit. As soon as she had a few teeth in her head, however, she started to bite hard biscuits and nuts. Pepel has always had a varied diet and although I buy plenty of fruit and nuts for her, her basic food is taken from my own plate. She likes to eat the same things that I eat and will drink anything that I drink.

I was quite surprised when on our arrival at Freetown again I found that Pepel had all her teeth showing. All except her four fangs, which developed much later. She now has a fine set of teeth and her fangs, two at the top and two at the bottom, are razor-sharp and needle-pointed. Her jaws are very strong, too, and she could quite easily sever a finger with one snap if she had the inclination. Although she is very placid, I should hate to be the one to offend her!

They had been building a new jetty at Pepel and we were supposed to be the first vessel to use it. However, another ship, the m.v. *Dalhana*, arrived there first, receiving all the glory. The captain of the *Dalhana* was presented with a silver tankard to mark the occasion and I can remember feeling very sorry for our own Captain Chinnery, who had tried to get the ship there on time. We missed, too, the lavish party laid on by the mining people for the first ship to arrive, but I made up for this by laying one on myself for the captain and officers and any shore people who could come. It was during this get together that I heard from one of the berthing masters that he had been told by the chief steward of the *Dalhana* that I was going to be sent to a new ship on our return to Holland.

When I told the captain about it he was surprised, but as he had heard nothing from the office he told me to treat it as a rumour. Knowing shipping companies as I do I could not help but believe that the rumour might be true. In some cases the ship's chandler knows where the ship is bound for before the captain! However, we were a long way from the U.K., so I did not worry too much at the time.

It was hotter than usual the last time in Pepel, so much so that I decided to sleep on deck, just outside my cabin. Pepel did not like this one bit and kept me awake all night until I thought it best to go inside, even if I just lay on top of the bedclothes. Pepel was happy about this and settled down to sleep, while I continued to lie awake thinking about my possible transfer. It was as well that I did go back into the cabin because a terrific storm broke loose and it poured down with rain for the rest of the night. Perhaps this was why Pepel was so anxious for me to vacate my deckchair.

The heat was so unbearable that we were all very glad when the time came for us to say cheerio once more to our friends at Pepel. As the *River Afton* slipped down river to Freetown I could not help feeling that it would be a long time before Pepel returned to her birthplace, if ever. She did not seem to mind, being too busy keeping out of the way of the relief crew who were preparing the ropes and wires ready for the regular crew to stow away for the sea voyage. Some of the crew pulled faces at her and while keeping at a safe distance she would reciprocate. People who are not used to this face-pulling spectacle usually ask if Pepel is savage. I jokingly tell them that she is only savage if they pull a face better than her!

# Chapter Six

Even as a baby Pepel was always a bit wary of humans until she got to know them, but it did not take her long to be friends with the radio officer, George Hargreaves. George was a friendly, quiet sort of chap and would not kill a fly intentionally. He never once tried to coax Pepel, but she always greeted him in her chattering way whenever he called on me. He would sometimes come for a quiet chat and a drink in the evenings when he was not on duty. I used to look forward to his visits because Pepel would sit on his knee to be cuddled most of the time, giving me a rest from the same duty. It was George who gave me the official news that I would be relieved at Ijmuiden. We had discussed the rumour many times during the trip back to Holland and he knew that I did not want to leave the ship I had only just settled down in. Poor old George was very apologetic about it when he broke the news.

Captain Chinnery told me the next morning. When I asked him where I was going he said that all he knew was that another chief steward would be relieving me at Ijmuiden and that I was to proceed on leave. Over a drink in the smoke room that evening Captain Chinnery told me how sorry he was that I had to go. It made me feel so humble and gave me a warm feeling inside, knowing that he was sincere. I had tried to help the Captain all I could since we both joined the *River Afton* together, knowing that it was his first command, and was very sorry to leave so many good friends and a grand old ship. I also began to worry about what was going to happen to my little Pepel if the Immigration officials would not allow me to take her with me to England. All sorts of ways of smuggling her through occurred to me, but all with the dreaded thought of being caught. I need not have worried about the Dutch authorities, however, because they passed her through without question. One of the officers even wanted to buy Pepel and offered me a hundred guilders for her (then worth ten pounds).

My own steward wept as he asked me to send for him when I went to my next ship. The night before we arrived, as I was packing my bags, the chief cook came and presented me with his recipe for 'Palm Chop' and his best wishes for the future. One or two of my own crew gave me their addresses and asked me to let them know if it was possible to come to my next ship. All this brought a great lump in my throat because I knew that I might never be able to see them again, let alone get them with me again. I thought as much of them as they did of me and since then I have often wished that I had such honest, hard working and sincere men working with me. The last person to bring his best wishes was my friend the tailor. He also gave me some more warm clothing for Pepel, who was very thankful for it a few days later. After handing over the department to the relieving chief steward, who was there on arrival, I said a final goodbye to everyone on board. My faithful boys carried my luggage to the waiting taxi as I went midships to say goodbye to Captain Chinnery. He gave me two envelopes which he told me not to open until I arrived home. I jokingly asked him if they contained secret orders. He replied, 'No, just something for you and our Pepel'.

The train journey from Ijmuiden to the Hook of Holland was an interesting one for Pepel, also for other travellers taking the same journey. They thought it was very amusing to see a real live monkey travelling with them, especially one dressed in a bright red jumper and a fur coat, with a scarf round her head to keep her ears warm! The dining car staff saw to it that Pepel did not go hungry. They brought her all kinds of fruit and nuts, sweets and chocolate, saucers of milk, etc. She repaid them for their kindness by picking up my beer bottle and drinking out of it, a thing she had quickly learned to do, for which I got the blame! It topped the bill as far as her Dutch audience were concerned just as it has done in many places all over the world since then.

No one at the Hook of Holland knew that a monkey had gone aboard the ferry for England. Pepel was safely asleep in my pocket and was still sound asleep when the baggage man had departed with a substantial tip after showing me to my cabin. After a good wash I retired for the night, or so I thought. During the night I had to leave the cabin to find a toilet and did not note the number of my cabin before I set out, carrying Pepel on my arm. The toilet was situated in another alleyway in what seemed to be a maze of alleyways and when I returned sleepily to my cabin I found myself in the wrong one! I became desperate after trying several doors, only to find other people snoring happily inside. Then a young girl came along in her dressing gown, obviously going to the same place I had come from.

She must have had a good laugh since, but at the time she nearly fainted on seeing a man standing in front of her, clad only in his underpants and with a monkey sitting on his arm! I felt ridiculous and breathed a sigh of relief when I managed to find my cabin to hide in. As I went to sleep again I could not help having a little chuckle over it, but also wondered what colour my face would be if I happened to bump into the same girl on arrival at Harwich. I made a mental note to apologise to her if I did because neither of us spoke on our alleyway encounter, we were too busy trying to vanish into thin air. It was all hustle and bustle on arrival at Harwich, but I soon found a porter for my baggage which made it easier to manage Pepel.

Pepel seemed to sense my concern as I approached the Immigration desk and kept under cover in my coat pocket while my papers were checked. Sweat was running down my face as I stood there for what seemed hours before the officer was satisfied and let me pass. So far so good, I thought, but I still had to go through Customs. The Customs officer went through the usual routine and asked me if I had anything to declare. I told him that I had two hundred cigarettes and a bottle of gin, then, as Pepel popped her little head out of my pocket, I added, 'Oh, also a small monkey'. The officer just smiled and said, 'What a beauty'! I walked on air to the train.

The journey to London seemed to take only five minutes. This was probably due to the fact that Pepel and I had got into the same compartment as an American family. The children saw to it that Pepel did not have time to become bored when she got fed up with looking through the window, and kept me busy answering questions about where Pepel came from and what I intended to do with her. The man asked his children if they would like a monkey like her. However, his wife said she had enough monkeys to look after, meaning the children.

Travelling across London to St. Pancras was even more of an experience for the taxi driver than for Pepel. He said it was the first time he had had a monkey for a passenger. It seemed a big joke to him and as we passed other taxis he shouted to the drivers and pointed to Pepel. She had the last laugh, however, because when I paid the fare at St. Pancras she snatched the half-crown tip out of the driver's hand, as if to show contempt for his joking about her. This caused a laugh among the porters who were eagerly waiting for another 'blood'.

Pepel was very well behaved on the train to Leicester, where we were to stay for a while. She looked through the window to watch the scenery and the strange animals in the fields, occasionally looking round to accept a titbit from some passenger and darting under my coat when the train went under a bridge or passed another train. Pepel has always had a fear of darkness and sudden movement. It is amazing how she is able to leap in any direction when she is startled. No one was there to meet us as the train pulled in at the London Road station because I had not warned my sister that we were coming. Pepel and I had to wait nearly half-an-hour for a taxi as so many passengers had got off the train at Leicester. However, the delay gave Pepel a chance to stretch her limbs and empty her bowels into a litter bin, providing some amusement for the people standing around also waiting for taxis.

I left Leicester at the age of fourteen when I ran away from home to 'join the Navy', although many years went by before I actually went to sea. As the years passed I lost contact with my birthplace and the people I grew up with, except for some members of my own family and even then it was by letter writing. During my absence vast changes had been made in the city I knew so well as a boy. Just how much it had changed I found out later, but meanwhile, as the taxi arrived at my sister's home in Clarendon Park, I wondered how she would take to Pepel, if at all!

Pepel was a bit dubious of Phyl at first. I think she was more than a little annoyed when the hugging session was on at the doorstep. Phyl, however, liked Pepel from the start and after a piece of apple had changed hands Pepel decided that Phyl was not such a bad sort after all. After she had eaten her apple Pepel went to sleep on my lap, no doubt feeling as exhausted as myself after over twenty-four hours at travelling. In no time at all my sister had a cup of tea made for me and we sat talking until Bob, my brother-in-law, came home from work. Phyl has always been very house-proud. No matter what time of the day it might be the house is always found to be spotless, not a speck of dust anywhere. Her kitchen, especially the cooker, is always immaculate. I have often wondered how she managed to cook such wonderful meals when the kitchen looked so unused. As I sat in her easy-chair, being careful with my cigarette ash, I hoped that Pepel would also appreciate her new temporary surroundings and not cause me any embarrassment.

I made a call to the office in Newcastle the next day and was told to relax for a month, then report for duty. At that time a month ashore would have killed me and when I told them this I was told to do a bit of gardening! However, after spending a week in Leicester, visiting my other relatives and losing myself in the city centre a few times, I began to get itchy feet. By that time Pepel was getting used to meeting people, more often than not in the pubs I frequented. The Craddock at Knighton seemed to be her favourite. It was there that I met George Neale, who happened to be a friend of my sister and family. Phyl had asked me to buy some fish for tea that morning, but meeting George put an end to our fish and chips that day. George is a pigeon man who likes his drop of beer at lunch time. I saw to it that his glass never became empty that day, while he told me all about his pigeons and how long he had known Phyl and Bob. If the police had been at the door at chucking-out time they would have had a good haul in George and myself. The former for being drunk in charge of a bicycle and the latter for being drunk in charge of a monkey and having no visible signs of fish on his person. I have often wondered what would have happened if Pepel had been in one of her drinking moods too!

# Chapter Seven

My sister Phyl has been very good to me in an almost motherly way since I can remember. She wanted me to stay with her longer, knowing that I had a month's leave. She also knew that I had to go, knowing me as she does for my wandering and my ever-present desire to move on to different pastures, meeting new friends on the way. As we stood on the platform waiting for the northbound train to come in I could not help thinking how nice it would have been if I had been able to settle down when I was younger, and to have a family around me like she has. It is times when those kind of thoughts go through my head that I am grateful for having Pepel.

I watched my sister fade into the distance as the train pulled out of Leicester to take Pepel and her master on to Sunderland, where later she was to become a well-known figure in the community. Being a great believer in 'the plan of life' I knew that there was a reason for going up north, apart from the business of going to sea. I didn't realise just how much a part Sunderland was to play in my life until a few months and many thousands of sea miles later. For the time being I was happy to be going there, where I intended to stay with friends for the rest of my leave.

Pepel and I had been lucky to get a seat in the same compartment as a fellow seaman who was travelling to join his ship at Newcastle. Apart from being excellent company on the train, he was also very helpful when we arrived at Newcastle, where Pepel and I had to change trains for the rest of the journey. There were no porters around and he was good enough to help me with my baggage over to the Sunderland train. Pepel was developing a liking for trains and hopped into the first open carriage door, looking back to see if I approved. As I sat waiting for the train to move she snuggled up beside me and went to sleep.

I knew that my friends would be waiting for us at Sunderland, but I was not quite prepared for the shock I received on arrival. Ray, who used to be one of my catering boys, stood there with his father, whom I noticed first. Ray had grown his hair long, almost to his shoulders, apparently because he was a member of some group. I did not recognise him at first, but when it dawned on me who this long-haired character was the shock left me speechless! When I first knew Ray he was a smart, clean lad who always had a neat head of hair and clean collar and tie. Standing on the station looking at him, even Pepel was a little afraid until he spoke, then she jumped up on his shoulder and urinated on him! In those days long-haired people were looked upon with distaste; nowadays, of course, they are part of the scene and taken for granted. It took me about a month to talk Ray out of his new hair style.

It was the first time I had seen my friends' new house on the housing estate at Pennywell. They had previously lived at Hendon, quite near the docks and not too far out of town. Although they now lived in a much healthier district I knew that the modern house and clean open spaces of Pennywell did not fully compensate them for the broken ties of Hendon. It was there that I met Joe Donkin and his son Ray and the rest of the family. Joe has been a dock worker since his army days and is also a well-known figure in Clubland. It seems that everyone in Sunderland knows Joe. Ray was only fifteen years old when his father asked me if I could give him a job on board the ship I was on at the time. He came on board as catering boy, rising to second cook and baker a couple of years later.

All the family took to Pepel with the exception of Betty Donkin, Ray's mother. She was absolutely petrified of her at first and would turn white whenever Pepel went near her. It made me feel very uncomfortable and I tried to keep Pepel away from Betty as much as possible until one day, as we were watching television, Pepel crept up to her and quietly lay on her lap. Betty went rigid with genuine fear, not daring to move. Usually, Pepel has no time for people who fear her, but to my amazement she put up her hand to Betty as if to say, 'Please be nice to me and don't be afraid'. From that day Betty lost her fear of Pepel and they became good friends, so much so that Lassie the pet mongrel became a little jealous for a while. It was always a wonderful sight to see Lassie and Pepel playing together or sitting in the window to watch the traffic go by. It was also funny to see Pepel run to Betty for comfort whenever I had cause to chastise her.

Pepel became very fond of Ray and I suspected that this was because of his long hair, not yet shorn. Her attachment to him became the cause of a little jealousy on my part. She became less obedient to me during her crush on Ray and after a while she actually preferred his company to mine. She would come to me while Ray was at work, but as soon as she heard his footsteps outside the door she would make little sobbing sounds and leap up into his arms as he came in. I began to wonder who owned Pepel in the end. I have never had a jealous nature so it was something of a new experience to feel annoyed at the obvious love Pepel had for Ray. This situation had its advantage, however, because when I had to leave Pepel for short periods in the future, I knew that she was with someone she liked and trusted. Ray played a big part in Pepel's life during the months ahead and my jealousy soon turned to gratitude. His first opportunity to 'monkey-sit' came when I went through to Newcastle to see what the office had in store for me. Pepel watched through the window as I went to catch the bus and it pleased me very much to see her still looking for me as the bus went by the house.

I arrived at the office of Hunting & Son with mixed feelings, worrying about my future as chief steward with the firm I had only been with a few months. I left the same office a very proud and happy man, to have lunch with Mr T. Shanks, the catering superintendent, and Mr Smith, his assistant. I had just been told that I had been chosen to become the Company's victualling officer on board their new flagship, the twenty-six thousand ton bulk carrier *Wearfield*. I was still in the clouds as we sat down to our lunch in a nearby Chinese restaurant. In between courses Mr Shanks gave me details of the new vessel and my position aboard her. The *Wearfield* was being fitted out with the intention of carrying a Chinese crew, but for the maiden voyage she would be carrying a British deck and catering crew. The ship would be ready in about a month if all went well with the fitting out and she would sail at the end of July if she passed her trials. Instead of being called chief steward I was to be known as the victualling officer and the new position carried with it the same privileges as other heads of departments. In the meantime I was to 'do a bit more gardening' until the trials, when I would be expected to help recruit a decent catering crew.

It was during those idle weeks that I came to love Sunderland and the people of Sunderland. They accepted Pepel as a 'reet grahnd lass' and Tiny, the 'man with the moonkee', almost became a Geordie himself!

One amusing incident happened at a social club I frequented at Hylton, on the outskirts of Sunderland. I took Pepel there a few times and she soon became a celebrity among the members, who would take little titbits to the club for her. They all loved her with the exception of one, it seems, because one evening as Pepel stood at the bar, she with her bottle and me with my glass, the chairman of the committee came over to me quietly and said, 'Please don't bring the monkey in again as it is offending one of the members'. I continued to go to the club occasionally when Ray could look after Pepel and a few visits later one of the members asked me why I did not take her to the club any more. I told him that I had been asked not to do so because of a complaint by one of the members. I have never seen a man look so angry as he shouted across the club, 'Oobaahdthamoonkeethen'? The word soon got around and so many people complained about how Pepel had been treated that the committee decided to hold a special meeting about it. The outcome of this was evident when I was invited to take Pepel on my next visit to the club. I should think that she is the only monkey in the world who has been barred and reinstated at a working men's club, but Pepel went even further than that a few months later. She was made an honorary member of the famous La Strada night club in Sunderland; but that comes later on in the story. Pepel being barred at the social club soon became a standing joke. For months after the incident club members used to shout to each other, 'Oo baahd tha moonkee, then?'

Pepel and I used to spend a lot of time in Mowbray Park, a very beautiful place of relaxation in the centre of Sunderland. Close by the park there is a pet shop called the Fish Bowl. I used to go in and purchase something trivial just to be able to go and have a talk to their pet monkey, Sammy. He was a dear little chap and I felt so sorry for him being kept in a cage. Although Sammy became very amorous towards Pepel she was never very interested in him, always being more concerned for the birds kept in cages nearby. She was never jealous when I gave Sammy a few raisins or peanuts, whenever I could draw his attention away from Pepel.

From the pet shop Pepel and I would go across the road to the park where, in later months, I used to go to sort out my problems or make decisions; but during this period all I had in mind was to give Pepel a bit of freedom in the sun in almost her natural surroundings. She used to like to watch the ducks swimming on the pond, much to the amusement of the kiddies, also there to watch the ducks. Whenever a duck plunged its head under the water for food Pepel would stand upright, with her hands crossed over her breast and an expression of amazement on her face, as if to say, 'What is that fool doing'?

I used to let Pepel off the lead in a secluded part of the park where there are little hills and grassy banks surrounded by trees and shrubs. I used to lie on the grass while Pepel scampered off to see what she could find among the shrubs. She would occasionally come running back to make sure I was still where she had left me and to let me know she was there she would jump on my face, until she felt the back of my hand!

One day Pepel went off on her usual hunting expedition while I had my little snooze on the grass. Suddenly I was awakened by a scream and, thinking that Pepel had bitten someone, I called out to her as I scrambled up to investigate. Much to my embarrassment she had merely stumbled into a courting couple on the bank and had scared the girl out of her wits when, thinking that the couple were fighting, she jumped on the man's back to try and separate them. No harm was done though and after I had apologised to them we all had a good laugh. Later, the young lovers became very good friends of mine and had their wedding reception at the restaurant I opened in Sunderland. Pepel enjoyed toasting the bride and groom with a glass of her favourite port wine.

As I was window gazing one day in Union Street, I noticed the reflection of an old lady sitting in a wheelchair behind me. She was staring at Pepel, who was also window gazing. Pepel seemed to feel that the woman was enraptured by her and quietly perched herself on the side of the wheelchair. The woman had no fear as she stroked Pepel behind the ears and talked quietly to her. Then she said to me, 'How much will you take for your little monkey?' She did not seem surprised when I told her that Pepel was not for sale. As she unwrapped a sweet for Pepel she told me that she used to own several monkeys and used to train them for a circus. She said that it was obvious that I thought a lot of Pepel and that her grooming was to my credit. Before we parted she offered me fifty pounds for her and promised that she would give Pepel a good home. When I refused her offer she gave me her name and address, 'just in case you change your mind'. Although I was impressed by the fifty pound offer I knew that no matter what happened in the future, I would never be able to part with my little friend.

From time to time Pepel and I would go to the docks to look at the ships arriving and sailing on the tide. We often went to the shipyard to see how the *Wearfield* was progressing. The maiden voyage would be delayed owing to the numerous strikes at the yard. It seems that she was doomed from the start, with one thing or another going wrong and the 'who should do which job' strikers holding up her sailing date. She was, however, taking shape and they were making a fine job of the accommodation. My quarters were very spacious. The smart cabin had a roomy bed, a settee, an armchair and a large wardrobe. There was also a beautiful desk with a telephone and a cabinet for books. A separate shower and toilet were situated next to the day room and everywhere was panelled with formica for easy cleaning. The same thing ashore would be called a self-tained flat. I became impatient for sailing day to arrive to be able to sample this luxury. The landlubber's life was becoming very boring for Pepel and me.

We spent the rest of our waiting period in going for walks along the river from Hylton to a place called Cox Green. This was a pleasant walk and not too long and tiring. Pepel especially liked it along the river because I was able to let her run free, to gambol into the fields at the side of the river bank. She played hide and seek with me, peeping at me from the hedgerow and disappearing into one bush to re-appear from another much further along the path. She enjoyed startling me by hiding in a clump of grass to wait for me to pass, then creeping up on me from behind to give me a nip on the ankle. Sometimes I saw where she was hiding but pretended not to so as not to spoil her fun. Her beautiful colours blend so well with the countryside that I doubt if she would be observed by anyone else. It is hard for me to detect her even in a room sometimes.

People often ask me if I am not afraid that Pepel might run away when I let her run loose. The only time Pepel thinks of running away is when she has done something wrong and knows that she is liable to be spanked. Even then she will not run very far but will wait until my temper has cooled, then creep back for forgiveness. I am also often asked how I managed to 'train' Pepel so well. My reply is that she has not been trained at all but has been brought up like a child. Since she was a baby Pepel has been cared for as human children should be cared for, with a lot of patience and understanding. She has always been nursed when she has needed affection. All monkeys love to be cuddled by their owners, a regular hugging session gives them a feeling of security, of being wanted. When she was very small I had to clean Pepel regularly when she messed herself. A lot of people would shudder at the thought of doing that chore, but it is not unlike changing a baby's nappy. It is certainly not as bad as having to clean up the mess made by a dog or cat. Pepel is not greedy, as most monkeys are, simply because she has always been fed at regular intervals with a varied diet. She never steals food and never stores food in her pouches because she knows that there is no need to do so.

Like most monkeys Pepel began to get very mischievous and, like some human children, very destructive. I cured her of the latter by giving her a spank whenever she decided to tear my books up or dismantle my ball-point pens. She knows that she is always spanked when she does anything wrong so she always thinks twice before committing an offence. Pepel knows too that the bigger the offence the harder the spanking is. On the other hand, Pepel has probably received more human affection than a lot of today's children ever hope to get, and in return her affection for me is such as I have never seen among humans. Hers is a true, loving affection given without motive, unlike the cupboard love of other animals. I am convinced that if I were to starve, Pepel would willingly go hungry with me.

I sometimes wake up early in the morning to find her sitting on my chest just staring at me, with utter devotion pouring out of her sparkling eyes.

I am often asked if Pepel makes any noise and, if so, what kind of noise. She rarely talks in public unless it is to speak to someone she knows. The only way to describe her vocabulary is that it is similar to that of a dumb child. She will make a different sound for a different meaning. Not like a cat or dog or any other animal. Pepel speaks a language and while I cannot speak in the same tongue I can understand what she is saying, just as she can understand what I am saying while being unable to speak English. Pepel seems even to be able to read my thoughts because she can always anticipate what I am about to say or do, just as I can anticipate her actions in every respect. That is how I am able to control her so well, especially in a crowded street when we are out walking. I am always very proud when people remark on her good behaviour.

When I tell people who enquire about Pepel's habits in the home that she uses the lavatory as we do, they can hardly believe me. It is quite true though and, although she will sometimes sit the wrong way round on the seat, she at least has the right idea. She does her early morning toilet on a sheet of newspaper, placed beside the bed each night. On odd occasions she will make an embarrassing slip, usually at the wrong moment when we are in company. I always carry some tissue in my pocket for this emergency. Taking everything into consideration, however, Pepel is as hygiene-conscious as it is possible for any monkey to be. She is certainly much cleaner in her habits than some human beings who are supposed to be more intelligent.

'Does she bite?' I am often asked this question, particularly by women who usually want to pick Pepel up to cuddle her. I always try to guard against this because, like any other monkey, Pepel will bite strangers who try to pick her up. Pepel is a model of good behaviour whenever she is in another person's house, or in public places. She expects other people to have the same respect when they visit us in our 'inner sanctum'. This could be my cabin on board ship, a hotel bedroom, the room given to me when I stay with friends or anywhere I sleep. Whenever anyone visits me in any of these places I have to take care that Pepel is out of reach when shaking hands with, or being embraced by, my visitor. I also have to see that persons other than close friends do not sit next to me on these visits, or Pepel will show her disapproval by giving them a nip on the ankle or on the arm. There is no real malice in this act, she hardly breaks the skin, but I try to stop her from doing this because it usually causes considerable alarm to the visitor. Although a monkey bite, if given in a vicious manner, can be very painful. Pepel would never give a vicious bite unless she did it to protect me. She is very placid normally and is very good with children and old people. She is also a very brave animal and will stand up to anyone or anything, big or small. Pepel sees red when she witnesses a child being chastised or a small dog being attacked by a bigger dog. She is also very gentle with babies; the only trouble is that she also likes to steal a baby's dummy and when the baby is crying over its lost dummy Pepel does not understand why. She sees no reason why she should not share this little bit of comfort.

# Chapter Eight

After completing her trials the *Wearfield* berthed alongside the Corporation Quay at Sunderland, where she was stored up for her maiden voyage. Storing a ship is always a hectic job at the best of times, but to receive stores on a brand new ship is even worse. A lot of mutual trust had to be exercised when accepting lorry-loads of provisions from drivers who had been waiting in a queue to deliver them. By the end of the first day I was signing delivery notes automatically and hoping that whatever I was signing for would be in the stores when we got to sea. It took the second steward and I a week to sort everything out. It also took me a month to compile stores lists from the dozens of invoices piled on my desk.

Sailing day soon arrived and I for one was very glad, in more ways than one. It seemed years since I left the *River Afton*, where I had everything ship-shape and I was looking forward to getting the *Wearfield* similarly organised. This can only be done at sea, where there are no distractions from ashore. A ship at sea becomes a little world by itself, with nothing to interfere with its smooth running until it enters port, where once again life becomes a bedlam.

Pepel had been looked after by Betty Donkin, while Ray, who had signed on as second cook and baker, was engaged with the rest of the catering crew in helping to store the ship. I had been far too busy to be able to look after her while this was going on, but I was very happy to see her as she was brought along the quay, on her way to the embarkation party being held in my cabin. What a party that was! It seemed to me that the whole population of Sunderland had congregated to give me a send-off. I never knew I had so many friends, who made sure that all the bottles were empty before the all ashore signal was given.

The send-off made us feel as if we were on a passenger liner. There were crowds of people on the quayside, shouting farewell and holding on to their end of the rolls of toilet paper being used as streamers. Then, with everything cast off, the tugs fussed around us as we slipped out of the harbour to make our way, as we thought, into the North Sea on the maiden voyage to the St. Lawrence Seaway, where we were to pick up a cargo of grain at Duluth, Minnesota, to be delivered to Gdansk, Poland. Unfortunately, after only eight hours at sea the *Wearfield* had to return to Sunderland owing to a fault in the engine room. There were quite a number of red faces as the vessel was re-berthed at the Corporation Quay and I knew that there would be plenty of leg-pulling when I took Pepel ashore that night.

It was quite funny to see the expressions on the faces of the friends we bumped into that night. 'That was a long trip', someone said, as Pepel and I walked into the lounge of the Argo Frigate for a pint. I told my fellow drinkers that

C

the *Wearfield* was jet-propelled. I also relieved the suffering of some of the crew, who had spent all their money before we left the previous day, by giving them a 'sub' to pay for their beer.

We sailed again the next day with a smaller crowd to see us off. Like the engineers, the rest of the ship's company kept their fingers crossed for a more successful attempt to reach Montreal. Although we made a good crossing, the *Wearfield* was to have plenty of teething trouble before the end of her maiden voyage.

Pepel soon settled down to her new environment and although she was not popular with some of the ship's company, she soon made friends with those who were not afraid to like her. There are some people who seem to be naturally afraid of monkeys, even to the extent of not being able to endure having a monkey near them. I have never been able to find out why this should be so, unless a monkey so much resembles the human being as to cause resentment. I do know, however, that the monkey recognises fear (or cowardice) in others. The only way to gain the confidence and respect of these fearless animals is by showing them that you want to be friendly without trying to touch them. Pepel will pull a face at someone she does not trust. She respects that person if he reciprocates and it almost always develops into a face-pulling contest, with the one who pulls the longest face being the winner.

There were smiles all round when the ship's carpenter made a miniature life-jacket for Pepel. She became very fond of him and loved him to play with her in the crew's messroom. He would pretend to sweep the deck while Pepel had a ride on the brush. When the play got a bit rough for Pepel she would seek refuge on his shoulder, only to come bouncing back down the brush handle when she had rested. Other members of the crew tried to play with

Pepel in a similar manner, but she preferred John Cormack, who, apart from being a good carpenter, was a genuine animal lover. She soon began to trust one or two of the others, however, and she would always be around at meal-times to steal a chicken drumstick or some other titbit from someone's plate.

Next door to my cabin was the officers' recreation room, mainly used by the cadets who played table tennis there. Although these lads were very considerate in regard to undue noise, the table-tennis balls pinging on the other side of my bulkhead became a source of annoyance to me when I was trying to do my figures. I never complained, realising that boys will be boys, but unknown to me I had an ally in Pepel. Long before we arrived at the St. Lawrence Seaway the pinging on my bulkhead suddenly ceased and I noticed that the recreation room was always deserted when I looked in. The reason for this became apparent when I asked the cadets why they had suddenly gone off their table-tennis? They told me that Pepel had chewed all the balls and until they could buy replacements the game was difficult to play! I noticed that they did more studying after this, so Pepel had probably done them a good turn inadvertently. I purchased a good supply of balls for them at Montreal, but by then they were kept too busy on deck to have much time for recreation, giving me time to catch up with my paper work before the pinging started again.

Pepel, now known as the 'Ship's Mascot', would watch the crew taking a dip in the swimming pool as we reached the warmer climate. They were all quite surprised when they did not have to coax her into the water. Once in, it was difficult to get her out. One day she refused to come out of the pool after everyone had got dressed, so we scuttled her by opening the outlet valve, leaving a very bewildered monkey at the bottom of an empty swimming pool.

On arrival at Montreal I was surprised to learn that my entire stock of fresh meat had to be placed under seal and any joints of cooked meat destroyed. The reason for this was because some of the carcasses in the cold store were not marked, making it impossible to identify their origin. The laws controlling the entry of live and dead stock into Canada and the U.S.A. are the strictest in the world. I was grateful for this later on when I was able to get a clean bill of health certificate for Pepel. Providing she does not return to her place of origin I am able to take her anywhere in the world without her having to go into quarantine.

After all official business had been dealt with the *Wearfield* proceeded on the slow journey through the Seaway to Duluth, Minnesota. The vessel had been built specially for the Seaway and was watched by crowds of interested people as she progressed through each lock. Pepel was playing out on deck most of the time, taking advantage of the warm sunshine and looking with interest at the beautiful scenery we were passing. Although I was not always on deck, I knew when we had reached another lock by hearing the crowds shouting, 'Look, they have a monkey on board'! Pepel came dashing into my cabin when she was frightened by a helicopter taking photographs of us as we neared Duluth. She must have thought it was some great vulture wanting to devour her, as she was very cautious about going out on deck again.

An American millionaire who was taking his small yacht through the Seaway always managed to squeeze into the lock under our stern. He came along to see Pepel while we waited for the lock gates to open. He thought she was very cute and took several photographs of her sitting precariously on the ship's rail, looking at her audience of sightseers who had gathered to watch the *Wearfield* go through her last lock. Suddenly, Pepel snatched the wealthy gentleman's expensive camera out of his hands, much to my horror and consternation. This caused the crowd to roar with laughter and struck the millionaire speechless for a moment; then he too had a good laugh, bringing from me a sigh of relief. I darted forward and retrieved his camera from Pepel, who had become a little agitated, realising that she had done something wrong. The millionaire photographer passed the incident off by saying that it would give him 'something to talk about'. I am often reminded of Pepel's camera snatching escapade when I see Boots the Chemists' very humorous advertisement called 'Where are you taking your pictures?' It is a depiction of a monkey in a cage at the zoo taking a photograph of an irate visitor who has just had his camera stolen by the monkey photographer!

# Chapter Nine

The State veterinary surgeon came on board with other port officials when the *Wearfield* berthed at Duluth. He told me that Pepel would have to go ashore to be examined and vaccinated and, as he had never handled a monkey before, he asked me if I would go along too. I told him that he would find it difficult if he tried to take her without me and anyway, as it was my first visit to Duluth, I might as well take advantage of having someone like him to show me around. He took Pepel and me in his patrol car to the animal clinic, where Pepel was examined by two surgeons and given the necessary treatment required by law in the U.S.A. We then had to go to the Health Department building where Pepel received her health certificate from the Minister of Health, who, poor man, was himself suffering from cancer of the throat. He said that it was always a great pleasure for him to be able to give a clean bill of health to another being, human or animal. This was the first time he had issued a certificate for a monkey and he hoped that she would give me a lot of pleasure for years to come. Pepel's certificate will always remind me of the man who looked with kindness on my healthy pet, while he suffered so much himself.

Pepel's certificate states that, 'One mango monkey, eight months old, green/gold color, female, was examined and found free from symptoms of contagious or communicable diseases and, to the best of my knowledge and belief, after reasonable investigation, has not been exposed to any such disease'. It is signed by the veterinary surgeon and the Minister of Health for Duluth, Minnesota. Looking at this official piece of paper I often wonder how many thousands of human beings there are who would give anything to be able to own a similar clean bill of health certificate.

The official business dispensed with, the State veterinary surgeon took Pepel and me on a tour of the city and treated us both to a slap-up meal in a restaurant where he was well known. Pepel, as usual, hogged all the limelight and we seemed to be in the restaurant for hours while everyone made a fuss of her. During the meal my new friend and guide asked me if I would like Pepel to appear on television. It seemed that before I had even a chance to give an answer, he had whipped us both round to the television studio and we were in front of the cameras before I managed to recover from this surprise! The studio technicians soon rigged up some scenery behind us as we stood under the powerful lights, Pepel doing her little trick of climbing up her lead to reach the beer bottle, while I perspired from nervousness and the heat of the lamps. Pepel seemed to know that she was expected to do her stuff and the studio staff were thrilled to have a real live monkey in their midst.

One of the television producers offered me five hundred dollars for Pepel. He was so intrigued by her cleanliness and friendliness and he thought that with a little training she would be useful in children's shows. It was very hard to refuse such an offer and I will admit that at the time it was a big temptation. However, the thought of going back to the ship without her was too much. I often think about that occasion and wonder whether Pepel would have become famous, had I left her in Duluth.

We also made a recording of an interview in the sound studio. This was later included in the rest of the day's news bulletins on the air. Listeners may have been amused to know that the twitterings coming from their radios were made by Pepel, caused by me giving her tail a little pinch in order to make her talk!

We were later taken to the Duluth Zoo, where Pepel visited the monkey house. She was taken there to see what her reaction would be to others of her race. She did what I have often seen children do when looking at Pepel – she just stared in amazement at the antics of the monkeys behind bars and when they chattered to her in her own language it seemed that she could not understand a word they were saying. The male monkeys thought they were being presented with another girl friend and became quite disappointed, and even outraged, when Pepel spurned their attempted love-making through the bars. Pepel thinks she is human and nothing would convince her otherwise. It was different when she saw the lions and tigers in their outside compound. She stood upright to pull all kinds of horrible faces at them, knowing that as they were behind bars they could not get at her. I dragged her away when she ventured too close in case those beautiful beasts got to thinking that monkey was on the menu!

That evening our new friend the vet took Pepel and me out to see his home on the outskirts of Duluth. I did not realise it then but what he really wanted to show off was his car. What a car! It would best be described as a bus, with every kind of amenity imaginable. Pepel liked to see the windows going up and down as if by magic. Everything seemed to work at the press of a button. I could see that its owner was very proud of it as he flicked specks of dust from the body with his handkerchief. He began to worry about Pepel's inspection of his pride and joy, thinking she might decide to leave something on the magnificent upholstery, so I thought it best for her to scamper about on the lawn nearby. Strangely enough, Pepel always enjoys a car ride and apart from her sea travels she has been driven many thousands of miles in cars since her visit to Duluth. She has never yet left anything on the upholstery.

While we were in town some people who originated from Manchester happened to bump into us and invited us back to their motel for a drink. The veterinary surgeon, who had given us so much of his time already, excused himself and left us with our new English-American friends. Although they had been in America for several years they had owned their motel for only a few months. They told me how hard they had had to work to get the place and how much harder they worked to maintain it. It was a credit to them and I could see how proud they were to be able to show off their successful labour to a fellow 'from home'.

We saw a lot of our motel friends before we left Duluth; in fact the ship swarmed with various friends the officers and crew had made while we stayed in Duluth. Such is life at sea. Friends are always made and seldom forgotten. Sometimes I think seamen would make better diplomats than those we pay to do the job. The seaman spends his life mixing with people from every walk of life in many different parts of the world. A lot of his time is spent in dockside pubs where life can be seen in the raw. Sometimes he is robbed in these places by people who prey on the gullible sailor, but most of the time he is meeting people who matter, ordinary folk who, for a while, are part of his life. Although he may never see these people again, the way they live and their views on life are impressed on his mind for ever.

The people of Duluth certainly gave me something to think about, especially when I was able to compare their hardworking, high living, happy way of life to the existence imposed upon the people of Poland. Some day I would like to return to Duluth with Pepel, who I am sure enjoyed her stay there as much as anyone. In the meantime the memory of our visit to such a wonderful place will always be a pleasant one to me.

# Chapter Ten

When we had left the Seaway and were ploughing through the Atlantic we soon felt the change of weather, especially the cold winds we were greeted by as we entered the North Sea. The crew, who had until then gone about their work half naked, soon decided to wear something warmer than the usual pair of shorts. By the time we reached the Baltic most of them were unrecognisable in their thick woollen jerseys and oilskins. The swimming pool became deserted and remained so for a number of weeks, much to the disgust of Pepel, who still ventured out on deck when my back was turned.

The captain's wife surprised me one day when she brought two little jumpers she had knitted for Pepel to my cabin. Until then she had been very reserved where Pepel was concerned and had never attempted to coax her except to throw sweets to her from a safe distance. I was very grateful for the jumpers, as Pepel had by then grown out of my old socks and the clothes made for her on the

*River Afton*. Her new jumpers were a perfect fit and Pepel lost no time in inspecting herself in front of the mirror with them on. She had learned to dress herself and it was very amusing to see her with her new jumpers. One of them being blue and the other pink, she could not decide which one to wear. She finished up by wearing both at the same time.

Everywhere was covered with snow when we arrived at Gdansk, where the ship was met by fussing tugs and a couple of motor launches containing armed soldiers and Customs officials. The reception committee which came on board as we berthed was something which had to be seen to be believed! I had never seen so many uniforms before as the officials who queued to come on board were wearing. I noticed that before the gangway was put out, armed soldiers were stationed about every twenty yards along the quayside, then when the officials had come on board, two more soldiers with automatic rifles stood guard at the bottom of the gangway. I had heard that these elaborate precautions were taken in Communist countries, but this was the first time I had witnessed them. No one was allowed to leave the ship until the entire ship's company had been interviewed by the police, who issued gangway permits. These permits allowed the holder to go down the gangway, where the permit had to be given up to the soldiers on guard, who then ticked off the holder's name in his book. We were told that we could not take British or American money ashore but we could change our money into zlotys on board. The official rate was sixty zlotys to the pound, but on the black market a pound was worth five-hundred zlotys. It did not matter how much Polish money we purchased on the ship, but we were not allowed to bring it back on board after taking it ashore. It was not until I had ventured ashore that I realised just why the Polish authorities had to be so cautious.

I did not go ashore the first evening we were there, as I was kept too busy entertaining the ship's agents and various port officials who had remained on board. Having been present when my stores were searched, these people lost no time in trying to scrounge as much as I was willing to part with. I had been told to be nice to them as much as possible so that 'things would run smoother' while we were there. Anything they fancied would be put on the entertaining sheet. I was thankful for this when I saw my stocks going down. The items they wanted most were things we take for granted in the western hemisphere, like razor blades, scented soap, toothpaste, ball-point pens and fruit. Razor blades topped the list because the small packets were easier to smuggle ashore. The more daring filled their briefcases with fruit for their children, who, I was told, only received this luxury at Christmas, apart from lemons, which could be obtained on a doctor's order.

I was reminded of this shortage of fruit in Poland when stores were delivered to the ship from the State chandler. I had ordered all kinds of fruit on my stores requisition but only received oranges and lemons. When the soldier on guard saw this rare commodity being unloaded on the quay he had the look of a child on his face. I offered him a couple of oranges and, to my surprise, he burst out crying. I felt awful standing there while this grown man cried, just because someone was being kind to him. He thanked me very much for the offer of fruit, but he only accepted one orange for his little girl. He told me in the best way he could that he would be punished if he was caught and as he would be searched by another soldier when he went off duty it would be better if he only took one orange.

I think I shall always remember Poland as a land of unsmiling people – people so oppressed that they just have no will to smile. The misery of their existence shows plainly on their faces as they go to work each day and as they return wearily to their dingy tenements at night. They seem to have nothing to look forward to, especially the women, who look with envy at a nylon shirt or a decent pair of shoes. They just could not buy nice things in Poland, even if they could afford to do so. Some of the crew took silk blouses and nylon stockings ashore, knowing they would fetch a good price on the black market.

When I took Pepel into town I was followed by a large gang of youths who wanted to buy pound notes. They were offering five to six hundred Zlotys for one pound. When I asked them why they were so desperate to buy British currency they told me that they wanted to get away from Poland and the people who smuggled them out of the country only accepted payment in sterling.

Pepel caused quite a sensation in the nightclub I took her to in Gdansk. I had been looking round the town with her and we had visited several shops without being able to buy anything, the simple reason being that although I had 'sold' several pound notes, I still had insufficient money to buy any one of the items on sale. I was grateful for this later, when I was able to purchase things of better quality and free of duty, from the State chandler's shop at the docks. It was while we were coming out of the last shop that I noticed someone waving to me from across the street. It turned out to be one of the agents, who invited Pepel and me to a club nearby.

I was quite surprised when I saw the luxurious appearance of the inside of the club. Until then I did not think they had such places in Communist countries. The agent told me, however, that the average person could only visit such a place once a month. I saw what he meant when I paid for the drinks, five hundred and eighty zlotys for one beer and a Russian-made brandy! At the official rate of exchange the round of drinks came to about nine pounds fifteen shillings, or approximately two weeks' wages for the average man in Poland! There are no bars as we know them in Poland, but in the cheaper restaurants, where only a kind of beef-burger with mashed potato is served, vodka can be drunk all night for a few zlotys.

Pepel was treated like a V.I.P. in the club, but I am sure that if she had known how much it was costing me she would not have been so keen on the brandy she was helping the agent to drink! Our table was surrounded by interested people who told me that they had never seen a monkey before. One person asked me if Pepel could speak. I replied that she had not been in Poland long enough to master the Polish language.

The thing that struck me most about the Poles was their open contempt for their own Communist Party and the Russians. I was told that the authorities knew about this but were not worried, knowing also that there were four soldiers to every civilian in the event of any real trouble. I was also told that the authorities knew it was useless to feed the people with propaganda, as they had tried to do in the past. Instead, they gave the people an anaesthetic, in the form of cheap vodka, to deaden the pain of their oppression. It seemed that the only thing the authorities worried about was how to prevent people from defecting to the West. This was being done on a large scale by those who could afford to pay to escape from such a miserable existence.

The depression of the Polish people, combined with the cold climate, soon spread to us on board the *Wearfield.* We were there for nearly three weeks, but by the end of the first week most of the crew were broke. By the end of the second week they had sold out of nylon stockings and merchandise originally bought for the purpose of selling at a profit in Poland. My turn to make a profit came towards the middle of the third week, when officers and crew alike started to use the ship's canteen more than usual, where they could purchase their beer on tick. The crew began to have parties in the messroom, much to the relief of the captain, who had by then become fed up with having to log and fine seamen who went adrift for days on end, after drinking the cheap local vodka. Some of them, however, stayed on board simply because of the severe cold and the miserable atmosphere ashore.

I stayed on board most of the time after the first week in Gdansk because I was afraid that Pepel might catch cold, despite her warm clothes. I had been ashore twice without her, leaving her in the care of Ray, the only other person she would be contented to stay with at that time. On both these occasions I felt lonely without her. I felt as though something was missing, as if I had forgotten something. She sobbed like a baby when I returned, so I knew she had missed me too.

I had plenty to do on board as I had been commissioned to do a 'Pictorial Record' of the *Wearfield's* maiden voyage. I had a heap of news cuttings and photographs to stick in the book, which kept me busy, when my little friend allowed me to get on with the job. She liked to play with the pieces I trimmed off the photographs, but sometimes she preferred to mess me about by picking up photographs I had sorted out to paste and, as some of them were already pasted, she got the rest of them sticky on the wrong side! This would go on until I lost my temper and put her to bed, from where she would peep at me as if thinking 'you bad-tempered old spoilsport'!

The day before we sailed the ship's agents were invited to have dinner on board. I tried to keep the menu as simple as possible so as not to cause embarrassment. They had a fruit cocktail, followed by minestrone, then a fish course, and the choice of several meats, including pork chops. Although there was a hot sweet and ice-cream on the menu, most of the guests preferred cheese and biscuits to finish up with. After the meal was over one of the agents told me that although Poland exported millions of tons of pork and dairy produce, the pork chop he had just eaten was as much as he was allowed to buy in a year, if he could afford to buy such a luxury! When I offered him a few chops to take home, he said that it was more than he dare do because, like everyone else, he was searched on leaving the docks and it would cost him his job, or worse still imprisonment, if he should be caught taking unauthorised goods ashore.

Our departure from Gdansk was just as dramatic as our arrival had been. No dockers were allowed near the ship and the docks were deserted, except for soldiers, police and a couple of port officials. When the ropes were cast off and the ship had left the quayside, the soldiers who had kept us company for three weeks, plus the extra men sent to make sure that no defectors got on board, piled into two trucks and were driven away. This still left us with our military escorts in the two launches and the pilot launch, which also had an armed soldier on board to make sure that its crew did not have any ideas about travelling with us.

The pilot was still on board when another launch came racing alongside with a message which no one could understand at first. A man on its deck was someone I seemed to recognise but I never thought any more about it until we received an urgent message on the radio. It seems that when my stores were put on board the bread had been supplied in metal trays which had not been returned. The man who was waving his arms about frantically was the official from the State chandler, who, the radio message said, would be in serious trouble if the bread trays were not returned. The Captain replied that they would be returned on our next trip to Gdansk, as he did not wish to delay a twenty-six thousand ton ship just for a few bread trays. As I had left the *Wearfield* before she returned to Poland, I sincerely hope that the Captain kept his promise, otherwise this poor man who was negligent over 'State Property' may even now be in Siberia.

# Chapter Eleven

It took a few days for me to shake off the gloom of Poland and I noticed that most of the crew were also more than usually quiet, keeping to their cabins when not on watch. The recreation room became a deserted place again until we reached Houston, Texas, when it was taken over by the immigration authorities as a clearing house and for the issue of landing passes.

We were all glad to be bound for the Gulf of Mexico, where we could enjoy the sunshine once more. The swimming pool had been given a good clean, ready to be filled for use as soon as the weather got a bit warmer. Pepel, too, needed the warmer weather; she had become very subdued since leaving Poland. This was probably due to the cold, damp air of the North Sea. I have always noticed that she becomes quiet when she is feeling the cold. At these times I try to keep her indoors as much as I can, without depriving her of the exercise she needs.

It was as we were steaming across the North Atlantic that things began to go wrong on the *Wearfield*. Machinery used for making drinking water out of sea water broke down several times, which put my department out of action until the engineers put matters right. Then we began to have power failures, usually while the bread was being baked, so that it had to be made all over again. All kinds of little things began to go wrong to upset the smooth running of the ship, which was also encountering bad weather. Later we were to forget all about these little inconveniences. In fact, we were very lucky to reach the safety of the land on the other side of the North Atlantic Ocean.

We were still several hundred miles out at sea when a raging storm broke loose, pounding the *Wearfield* as though she were a mere piece of cork. We seemed to be bobbing up and down in the same hole for hours, until a fresh mighty gust picked us up again to drop us into another hole with a great shudder, where we bobbed up and down until the next gust came along to move us to the next hole. I had been in very rough weather many times before without worrying about it too much, but the antics of the *Wearfield* frightened me as I had never been frightened before. Pepel was very frightened too, and was sick for days. She could not understand what it was all about and I was too busy helping in the galley to be able to comfort her. I had to tie her to the settee in my cabin while I was away for fear that she might get out on deck and be washed away. I would return to the cabin as often as possible to let her see that I had not deserted her. Each time I went I had the most awful mess to clean up. I did not mind this, knowing that she could not control herself through being so scared.

The weather became worse and the sea was relentless. Each wave broke over us as if trying to smash us to the very bottom of the Atlantic. Some of the waves were at least eighty feet high and it seemed as though we were going through a wall of water most of the time. Just as I was thinking that the ship would not be able to stand much more of this punishment, the word went round that something was seriously wrong in one of the holds. A terrific crashing noise could be heard with each roll of the ship and it was thought to be something that had broken loose with the constant pounding on the ship's hull.

The Captain decided to alter course and put into the nearest port before more damage was done. The vessel heaved and shuddered her way to Nova Scotia, where she limped into the port of Halifax with her holds full of water. Strangely enough, it was not the *Wearfield* but Pepel who was splashed on the front page of the Halifax *Mail Star* the next day.

'MONKEY IS SHIP'S MASCOT'

'A nine-month-old monkey, which was picked up in a tree in West Africa, and has since sailed from the Baltic to the Gulf of Mexico, is in Halifax on an emergency stop-over.

'The name of the monkey is Pepel, the gender is feminine, her job – mascot aboard the new 26,000 ton British ship, Bulk Carrier *Wearfield*. The ship is now at Pier Nine for emergency repairs. She is expected to sail later today for a load waiting in Houston, Texas, and meant for delivery in Gdynia, Poland.

'The ship was some 600 miles out in the Atlantic from Halifax, when she was struck by heavy seas and winds up to 70 mph. A loading partition in the holds gave way and cracked. The danger was that the ship's hull would be damaged. Captain Marshall Friskney decided to make for Halifax and have the damaged partition removed.

'Meanwhile making the most of the first opportunity in

her life to caper in the snow, Pepel was out and about on deck yesterday, dressed in a warm, red jacket made out of an old sock by the ship's carpenter, John Cormack'.

The article then went on to say how I got Pepel and told of her visit to Duluth. There were two photographs of her, one showing her smoking a cigarette as she looked through a porthole, the other as she lay in the crook of my arm while being admired by the newspaper reporters.

After the newspaper report we had scores of people visiting the ship to see Pepel. One woman brought her entire family, loaded with fruit, nuts, gum and candy to feed her with. A party of school children came to see her and take photographs of her. Another wealthy woman brought her son to see Pepel and became very indignant when I refused her offer of a good home for Pepel, 'who would make a lovely playmate for the boy'. It seems that her son already had a collection of pets and he had hoped to add my pet to it. Pepel proved to them that money cannot always buy everything when I allowed the boy to walk her down the gangway. She pulled the lead out of his hand and came running back to me sobbing, as if to say, 'I would not go without you'!

I believe that we all thanked God for bringing us through the storm to the safety of land, but I was also pleased that the land happened to be Nova Scotia. Although I had never been there before, I had heard a lot about the hospitable Nova Scotians. It was funny to see Pepel running and jumping in the snow as we walked ashore that evening. I tried carrying her, but she insisted on sampling the feel of the crisp snow under her feet and I had to quicken my pace to keep up with her. She was well wrapped up in her jumpers, making a colourful sight against the background of thick snowflakes which were already covering our tracks on the soft carpet of snow.

Although I too was wearing a thick scarf and an overcoat over my uniform, I was grateful for the warmth of the bar I stamped my feet into on the way into town. The room I carried Pepel into was full of cigar-smoking American Marines, who were on a courtesy visit to the port. They turned out to be a very friendly bunch of lads who soon wanted an introduction to Pepel, who already had her beady eyes on the money left carelessly on the tables by these trusting Americans! Pepel has always been very fond of collecting and playing with money, so it did not surprise me when she went round all the tables, systematically relieving the Americans first of their dollar notes, then of as many of the loose coins as she could manage to stuff into her pouches!

The good-natured Marines did not mind this audacious robbery and even thought it was 'cute'. To my embarrassment, they asked me how long it had taken me to teach her how to steal? I rejoined by telling them that she was not stealing, but only borrowing. Pepel seemed to know that they all loved her and what a laugh we all had when, to show her appreciation, she went to each one in turn and sat on their shoulders as she looked through their hair and, in some cases, their beards! They all thought it was great and invited me to take Pepel on board their ship, an invitation I refused because I wanted to see a bit of Halifax before we left the next day.

Leaving my new American friends, I tucked Pepel under my arm and went in search of a shop where I could buy some postcards of Halifax and get Pepel a new lead, as the one I had made on the *River Afton* was now looking a bit shabby. Pepel wears a chain round her waist as we wear a belt. The chain has a special ring on the back where the lead is fixed when I take her out. Some people are under the impression that the chain is somehow fastened to her tail, which is now two feet seven inches long. However, when I show them how it is fixed and assure them that it is in no way uncomfortable to Pepel, they are satisfied that there is no case for the R.S.P.C.A.

After purchasing Pepel's lead and a few postcards, I had a look round the city centre. I was disappointed in Halifax itself, imagining a far bigger place than I found it to be. Probably if I had had more time I would have been more impressed, or perhaps it was because I was kept busy by people admiring Pepel that I missed the better sights of Halifax. I could not fail to like the people though because, although they seemed to be a little reserved, the people who spoke to me were very kind and their admiration of Pepel melted any ice there might have been.

With fingers crossed, we resumed our journey to the Gulf of Mexico the next day, leaving behind the damaged partition on a quay that was almost deserted except for the men who let go the ropes and a little Indian boy who had come to see Pepel. She sat on the ship's rail until the pilot was dropped, when she scampered back into the warmth of my cabin.

It did not seem too long before we sailed into the sunshine and the swimming pool became an attraction again. We had no further trouble in the engine-room, much to the relief of the engineers, who were now able to relax a bit more on deck. They seemed like strangers to everyone as they emerged wearing clean uniforms, instead of the greasy boilersuits they had worn most of the trip. The electrician we had on board was very fond of Pepel. He was always playing with her when he had the time. She had missed him while he had been kept busy and lost no time searching through his hair as he lay sunbathing near the swimming pool. I, too, was delighted to see him back on deck, because I should know where Pepel was while he was around.

So the days passed until we reached the Gulf of Mexico and Houston, the home of the oil millionaires, but where a spade is a spade, no matter how many dollars you possess.

# Chapter Twelve

There are a lot of jokes about Texans in circulation, most of them relating to their bigness of head rather than their stature. Some of the jokes have been good-naturedly put around by the Texans themselves; a little self-ridicule is nothing to these amiable folk, who are generally big enough to stand a bit of leg pulling, anyway. Other, more crude legends are spread about by people who are obviously jealous of the Texans' wealth and masculinity.

Whatever else Texans may be full of, they are certainly not lacking in the milk of human kindness, as I found out on my visit. Yes, everything is big in Texas, but their biggest 'bigness' is the instant friendship they offer to all who go into their midst. Money does not come into it, rich and not so rich rub shoulders and know each other by their first names. A spade is a spade and, depending where he digs the hole, one man can be as rich or as poor as his neighbour overnight.

I had hardly stepped off the *Wearfield* when I found my

first friend in Houston. He turned out to be the taxi driver. who went out of his way to be friendly when he took Pepel and me on a shopping spree in town. He took great pains to take me to the stores where he knew I should find the things I needed. I began to worry about the fare when I was held up by people wanting to admire Pepel, knowing that the taxi was waiting outside the store. The driver also seemed to be taking us on a roundabout route, while he described the various places of interest to me. When I told him that I could do the rest of my shopping on foot he would not hear of it; then, when my shopping was finished, he asked if I would like to go to his home for a little refreshment. When I worriedly asked him how much I owed for the taxi, which I had been in the whole afternoon, he said, 'Aw shucks, jus bring tha liddle monk ta see ma kids, an ya don owe me nuthin!' (I hope I have spelt the accent right.)

After phoning his wife to tell her to lay two extra places for dinner, this overwhelmingly big-hearted, ordinary Texan drove Pepel and me to his home for a meal I shall never forget.

He drove out into the country not too far out of town, where he and his wife and family lived in a wooden house which was raised from the ground on stilts and had a large patio in front. His wife, who greeted us as we stepped out of the taxi, did not seem very surprised when she saw Pepel sitting in the crook of my arm. She explained her lack ot surprise by telling me that she never knew who or what her husband would bring home from one day to the next. She had got so used to it that nothing surprised her any more.

Judging by the meal she later served, it seemed that she always kept a well-stocked larder for the surprise guests her husband took home. We sat down to what she called 'smothered chicken', a whole chicken cooked in a casserole, covered with a bread stuffing. I have often tried to cook a similar dish without success, or at least it has never tasted so delicious as the Houston original. The chicken was followed by thick steaks cooked in a thick chili sauce, served with tossed fresh salad, chopped fresh red peppers, sweet corn and baked potatoes. After wading through that lot I was unable to do justice to the apple pie and cream, etc. Pepel, who had never seen so much food all at one time, also enjoyed her meal and even managed a little of the pie, after which we all sat on the patio and smoked while Pepel played hide and seek with the children under the house.

It does not take very much exertion for me to perspire and, as the chilis I had eaten were taking their effect, I was grateful that Pepel had someone else to perform her physical activities with. I still felt a little strange in the company of my new friends, who were doing their best to make me feel at home. The main topic of conversation, of course, was Pepel, by then a favourite with what seemed to be the whole neighbourhood, who had gathered as if by magic to watch her jump in and out of the taxi. Her audience laughed their heads off when she found out how to press the button, to make a noise with the horn. Pepel thought it was great fun and played to the audience.

Observing the gaily decorated Christmas tree standing elegantly in a corner of the patio, reminded me that it would soon be the one day in the year when most of the world dropped tools to celebrate the birth of Christ. Politicians would stop slandering each other to send each other 'goodwill' greetings. The armies engaged in fighting the politicians' squabbles in various parts of the globe would cease fire, pausing to give each side a rest from killing and another day to live. Church dignitaries would become more reverent. People would pray harder, some openly, some in secret. Bells would ring out their message, choirboys would sing it all over the world – the message of love, 'peace and goodwill to all men'.

What a pantomime this world of ours has become and, if we would only admit it, what hypocrites we are! Posters plastered all over England ask people to 'help fight the war on want'! Why anyone in the world should want is beyond my comprehension, except perhaps it is the greedy people of the world who prefer to make weapons of war, to guard the rotting grain in their storehouses, rather than give it to others who are hungry. I have seen in my travels whole granaries full of wheat sprouting and taking root in the grain already decayed at the bottom. I have seen warehouses full of sugar smelling like distilleries, with sugar oozing like treacle over the floor. I have also witnessed in England whole fields of vegetables being ploughed into the ground because the 'right price' had not been made for it. I have seen tons of fresh fish being dumped back into the sea for the same reason.

Although the chosen moment is sometimes irritating, does it not give us a wonderful feeling of neighbourliness when the woman next door is not afraid to come and borrow a cupful of sugar, or any other commodity?

I was saying goodbye to my taxi-driver friend and his family when this actually happened. I had, while enjoying the surroundings from the patio of their lovely home, been secretly comparing the wonderful secure life of Houston to the miserable existence, described in Chapter Ten, of the people who are forced by birth and their Russian overlords to live in the hungry coldness of Poland. Judging by what I saw there, I can imagine the same kind of existence in other Communist countries. I also thought about the soldier who wept when I gave him an orange for his child, the man who dare not accept another for himself for fear of being caught. I wondered how many women in Poland could borrow a cupful of sugar, like the woman in Houston. I thought, too, of the kids in Gdansk at Christmas as they queued, each receiving an orange from kind 'Father State'!

Monkeys are supposed to be the greediest of animals, forced by nature to grab all they can while the going is good. Wild monkeys will also destroy what they cannot eat or stuff it into their pouches, in case it is of some use to their foe. Pepel is only slightly greedy when it comes to the things she likes most, like the cocktail cherries she consumes by the dozen when she gets the chance. Like all monkeys, she will not allow anyone to take anything from her, but, unlike most of her race, she is willing to share whatever she has with her friends, human or animal. Pepel proved this to be true one day in a Houston park. While I sat surveying the scenery Pepel busied herself with the task of eating an ice-cream, holding it in one hand just as we do, licking the ice as it melted and ran down the edge of the cornet. She was thus engrossed when a young lady, who was pushing a pram, stopped to stare in amazement at this unusual sight. Then, to her horror, Pepel jumped on to the pram to give the baby her ice-cream, at the same time stealing the baby's dummy-tit to put into her own mouth! After the initial shock of this kind, but audacious act, the young lady and other spectators relaxed to laugh at this most unusual and friendly monkey.

Pepel made many friends in Houston, particularly the children who, like the children in Halifax, came to visit her on board the ship. Like Pepel, those children are almost five years older now, teenagers who have probably forgotten the cheeky monkey who came into their midst. Pepel was all legs and fluffy hair then. Now she is a very beautiful teenager, too, who has since had many experiences all over the world, some happy, some almost tragic. I doubt, however, if Pepel will ever forget the people she met in friendly Houston.

I have lost the address of my taxi-driver friend, but if he happens to read this book I hope that he will get in touch with me. I would like him and his family and friends to know that I have never been able to forget the wonderful, homely, big-hearted Texans, who made the visit of Pepel to Houston a most memorable one.

It was hard for all on board to leave our friends as we sailed from Houston back to Europe. My cabin looked like a greengrocery, with fruit, nuts and all kinds of gifts which had been sent on board for Pepel from her admirers ashore. Nothing would please me more than to be able to return to Houston one day to show them how beautiful Pepel has grown since they knew her as a baby.

# Chapter Thirteen

We were on our way to Rotterdam when Christmas 1964 arrived. It was Pepel's first Yuletide experience and she seemed to be as excited as the rest of us when she saw the trimmings going up. She watched with great interest as the Christmas tree, given to us at Houston, was being decorated. She was not so keen on the Father Christmas made by one of the deck officers out of a red ensign and cotton wool, complete with a pair of Texas boots! The dummy stood in a corner of the officers' saloon and looked, as Pepel thought, quite life-like. She kept creeping round to the saloon from my cabin to peep at it, while keeping at a safe distance, from all angles. She would venture so far towards our jovial-looking Santa Claus, only to retreat at the point of touching his boots. Her antics caused quite a laugh, which was just as well, because although we tried to make Christmas a happy one for all on board, bad weather and watch-keeping dampened what little festive spirit there was on the ship.

Pepel received Christmas presents from several members of the ship's company, including some dresses from the captain's wife. Although she was obviously being nice to Pepel, I was secretly aghast when she gave me the dresses for her. I know that some people think it is very clever and amusing to dress up a monkey. I just think it is foolish and, apart from making her wear a warm jumper when it is really cold, she has never been made to look like a clown in fancy dresses. In any case, Pepel has such a fine natural coat, it would be a pity to hide it with man-made clothing.

The gift that pleased me most was the miniature life-jacket Pepel received from the carpenter, John Cormack. He had made it himself with great attention to detail, complete with the printed words, 'M.O.T. One Monkey'. I shall always remember John for his kindness to Pepel and would like him to know that Pepel's life-jacket is with the rest of her treasures in her toy box.

Although I did not know it then, Christmas 1964 was to be my last Christmas at sea for three years. The sea had been my career for fourteen years and at one time I could never imagine myself doing anything else. The wanderer's life had always appealed to me, until Pepel arrived on the scene. When I first realised that I could never part with her, I began to have ideas about settling down, so that I could give her a proper home and a run in the fields every day. Although she was never actually caged up, my duties at sea made it impossible for me to give her the freedom she would have liked. Also in the back of my mind lurked the fear that she might one day be lost over the side. This thought sent ice-cold shivers up and down my spine.

I knew that there would be no problem in finding suitable employment ashore because, apart from being a chief steward, I am also a qualified chef de cuisine. The only trouble was, I thought, whether I should be able to look after Pepel any better, or at all, if I obtained a chef's position in a hotel or restaurant, where long and awkward hours are worked.

Having lived a cosmopolitan life since I was fourteen years of age, I had no real home anywhere. I brushed aside the thought of going to live with or near my family in Leicester because of the unhappy memories of my youth in Leicester. I also dismissed the idea of living permanently with the many friends who had offered me a home when I was ready to settle down. I did not want to impose on their kindness.

To look after Pepel properly I had to be my own boss and have my own home, where Pepel could do as she wished without being an inconvenience to anyone, or where I would not have to get permission from anyone for her to live with me.

The more I thought about it, the better the idea of having a restaurant of my own became. I did not have much money, but I thought the little money I had would start me off; my hands and plenty of hard work would do the rest. It was a gamble and although I am not a gambling man, my experiences during the years at sea taught me that life itself is a gamble and, although one does not always come in first, one has to be in the race in order to win it. I had never had a business of my own before, although I had often thought that my talents, used till then for making money for other people, could be used to better myself in that direction.

Having decided in favour of a restaurant of my own, my next problem was the location of such a place. This was solved one evening at a Christmas party on board the *Wearfield*. I was entertaining the chief cook and the second cook and baker (Ray), who were the first to arrive at the party I gave to the catering staff in appreciation for the extra work they had done during the Christmas period. Pepel, as usual, was also present to help me entertain the guests, reminding me when anyone needed another drink by draining their glasses for them. No one could ever be bored at a party which included Pepel.

I was discussing my intention of becoming a restaurateur and both thought that it was a good idea. When I told them that I was undecided about where to establish a café or restaurant the chief cook said, 'Why not open a place in Sunderland, where you are already well known and have friends who may be able to help you in the venture?' Ray also said that a restaurant in Sunderland might be a good thing because, to his knowledge, there were only about three such places there and these could do with a bit of competition. It needed thinking about and there was still plenty of time, but I liked the idea and did think about it for the rest of the voyage.

New Year's Eve was celebrated at Rotterdam, where the *Wearfield* dropped her anchor before proceeding to the Kiel Canal and continuing her voyage to Gdansk. The crew had become unsettled between Christmas and their arrival at Rotterdam. Several of them jumped ship when the captain refused them leave to go home for the New Year celebrations. They did not see why they should not be able to do this with the ship remaining at anchor for three days. The captain knew, as captains do, that if he allowed them to go home there was more than a possibility that some or all of them would not be back on board to get the ship away again. Knowing seamen as I do, I knew that his decision was right and proper, although I felt sorry for some of the seamen who really wanted to see their wives and families at this festive time.

I was more fortunate in that I was able to leave the ship

at Rotterdam by a mutual agreement between the master and myself. I had made up my mind to have a go at the restaurant business and, having made that decision, I knew that it was a case of now or never. Making a break from my career at sea was going to be tough in any case, leaving it any longer would only make me hesitate and possibly change my mind. It was like jumping off a cliff and hoping that there was water at the bottom. I simply had to take the plunge. The knowledge that Pepel's future happiness was at stake also made me realise that once I took the plunge I would have to do all in my power to keep my head above water for her sake.

I must confess that I had a great feeling of insecurity when I stood, holding Pepel, in the launch which was bobbing up and down alongside the mighty hull of the *Wearfield*. Looking back at her as the launch was whisking us towards the shore, and an unknown future, I knew that there would be no turning back. I did not know whether the feeling I had in my stomach was relief or uncertainty.

After being paid off at the British Consul's office in Rotterdam, I had all day to wait for the night ferry to Harwich. The weather was very cold and the streets were covered with slush, which made it difficult to walk in the ordinary pair of shoes I was wearing. I thought then that what I really needed was a pair of good Texan boots! I had toyed with the idea of staying in Rotterdam for the week-end and visiting some friends in Schiedam, but as the weather was so cold I did not think Pepel would appreciate it.

Pepel and I made our way to the Hook of Holland and sat in the station buffet for the rest of the day. I for one did not mind because, unlike the British railways, the station buffets on the Continent are always warm and, in Holland, have the friendly, almost family, atmosphere which is typical of the Dutch people.

I was almost ready to go to the ferry when I saw two seamen off the *Wearfield* come walking into the buffet, complete with baggage. They had jumped ship and were going to try to get back into England with their American landing permits, their discharge books and identity cards or British passports still being on board the ship they had just deserted.

There was an even bigger shock waiting for me when I went about the ferry. Ray, whose home I was going to in Sunderland, was also on board! He too had jumped the ship and intended to travel on the ferry without a ticket, as he had no money to buy one. He had about a hundred and thirty pounds due to him from the *Wearfield*, but I knew that he would have some difficulty in obtaining that now that he had deserted ship. I got a ticket for him, also a berth in my cabin, but, like the others who had jumped, he had only his American landing permit as a means of identification when he arrived at Harwich.

I had no worries about Pepel's re-entry into England the next day, due to the fact that she had her American health certificate. I did worry about Ray though and hoped that if there was any trouble with the immigration officials I would be able to help smooth things out. I need not have worried, however, because Ray and the two other seamen were checked through with a little grumbling by the official, who wanted to know if there had been a mutiny of some kind on the *Wearfield*. It seems that the same official had been on duty when other deserting seamen had come through, all with American permits! I do not think they would find it as easy to come into the country now with the new and strict regulations in force.

# Chapter Fourteen

It was with a sigh of relief that I settled into a corner seat on the London-bound train standing at the little Harwich station. Whenever I travel on a train I like to get a corner seat so that Pepel can look through the window. I can also shield her from the heavy hands of some of the inconsiderate people who think she is there solely to be patted and poked. Fortunately for Pepel, there are only a few of this type of person. Most people are content just to admire her without wanting to maul her about.

Ray, who was sitting opposite to me, was thoughtfully looking through the carriage window as the train pulled out of the station. He was, no doubt, thinking about what he had committed himself to by jumping ship. His Merchant Navy discharge book would be sent by the Master of the ship to the Registrar of Shipping, duly marked, 'Voyage not Completed'. This would be a bad mark against him when he subsequently tried to get another ship. He would also forfeit all his pay, which I have said before amounted to about a hundred and thirty pounds. The fact that he was not the only one to desert was no consolation, but it did prove that the *Wearfield* was far from being a happy ship.

As things turned out, both Ray and I were destined to work very hard together during the next three years. The worries I had as a chief steward were to be very small indeed, compared with worrying about my own business. Neither of us realised, as we arrived in Sunderland, that within a month we should be running our own restaurant and, funnily enough, would be worrying where the next meal was coming from.

The first week in Sunderland was spent in looking round for suitable premises for the restaurant. With the limited capital I had, I needed a place which would need only the minimum of money spent on it by way of decorating, etc. After looking over several properties we finally found what we thought would be the right spot, a building on the corner of a block of shops, quite near to several housing estates and not too far out of town in a suburb known as Pallion. To me, the place was perfect for the type of restaurant I had in mind, a place that had a cosy, intimate atmosphere where a good class clientele could be cultivated who would be able to enjoy high-class English and Continental food at moderate prices. Although I knew it was going to be a gamble, my enthusiasm increased after negotiating for and obtaining a five-year lease for the premises soon to be known as the 'Restaurant Las Palmas'.

Las Palmas, in the Canary Islands, has always stuck in my mind, being the first place abroad I visited when I was 'Captain's Tiger' on board a passenger liner which, for obvious reasons, I will not name. The incident which impressed Las Palmas on my mind occurred when the ship called there for bunkers and fresh provisions before continuing her voyage to South Africa.

The captain and chief engineer went ashore with the ship's agent, supposedly on ship's business. The vessel was due to sail at eleven o'clock the same evening, keeping to the tight schedule of a passenger liner on that service.

We had about a thousand passengers on board, most of whom went ashore during the eight hours we were there. They all drifted back before sailing time and lined the decks nearest to the quayside, waiting to see the lights of Las Palmas disappear as the ship sailed. At eleven o'clock nothing happened. Everyone waited but still nothing happened at eleven forty-five. There was no sign of the ship sailing at midnight. The gangway was still out and the master at arms at the foot of the gangway seemed to be looking for something to come from the direction of the town.

The chief officer, worriedly pacing the bridge with the equally perturbed pilot, decided to blow the ship's whistle for two people who were not yet on board – the captain and chief engineer!

Soon after, a taxi came screaming to a halt on the jetty. The chief engineer stepped out of the front door, staggered to the rear of the taxi and dragged the captain out of the back seat, ably assisted by the master at arms and two quartermasters. They were both quite drunk, especially the captain, who had to be half dragged, half carried to the gangway.

None of the passengers realised that the two drunks being helped up the gangway were the captain and chief engineer, as both were wearing civilian clothes. A kindly old lady, going out to see her son, a missionary in Kenya, said to the captain as he was half way up the gangway, 'Come on, Pop, you nearly missed the boat'! The captain replied by giving her a glare and an inverted 'victory V' salute. I was horrified at this behaviour because I knew him to be normally a very fine and considerate gentleman who would not dream of offending anyone. I was glad for his sake that the incident, very amusing to us seamen, was also passed off good humouredly by the old lady and the rest of the people who witnessed it. My job as the captain's steward meant that it was I who finally put the 'old man' to bed to sleep off the effect of the Las Palmas aniset he had consumed. After performing this task I quietly put it around the ship that the captain was not drunk but had become ill while he was ashore.

Pepel was not born then and indeed at that time I could not have believed that one day I would be the owner of a monkey and a restaurant named 'Las Palmas'.

Ray and I spent the first week of occupancy on the task of painting and decorating the interior of the place, comprising a large shop at the front with another large room and kitchen at the rear on the ground floor. The flat, which contained two large rooms and a bathroom, was situated on the second floor and had a separate entrance at the side of the building. There was a good-sized yard and a garage at the rear of the building. The garage was later turned into a store and preparation room. There was plenty of room for expansion should the occasion arise, as it did later on.

As far as Pepel was concerned, the yard became ideal for her to play in when the weather permitted her to do so. When we had been at the restaurant a while Pepel got used to her yard and enjoyed sitting on the wall to watch people passing by. She also attracted attention and people became curious to know the owners of the new restaurant where the monkey sat on the wall.

I was informed by various people that others had tried to establish 'the café business' in Pallion without success. Some of them even went as far as giving us six months before we went out of business. I knew that if we did go out of business it would happen in less time than they anticipated. We had to get established within three months or go bust! In actual fact, the first three months were very worrying for me. Every penny I possessed had been sunk into the venture, besides other money I had borrowed from a friend, who became a silent partner. It was his money I was really concerned about, specially when our first day's takings amounted to just ninepence!

As we could not afford expensive kitchen equipment, everything was done by hand to start off with. Potatoes were peeled and chipped by hand and fried in an ordinary egg pan. Ray and I did the cooking while his mother served. After a while, when we began to get busy with the workers' three-course luncheons we had started to do, we had to get a part-time waitress to assist Peggy. Our luncheons soon became the talk of the town and we began to get people coming from shops and offices in the town centre, where there were restaurants and cafés close by. This gave us encouragement and we were soon working eighteen hours a day, opening seven days a week, including public holidays.

Our evening trade began to pick up after about four months. We had Indian curries on the menu as well as English and Spanish dishes. Curry lovers came from all over the North-East to sample our cuisine and our popularity increased so much that within a few months of opening, the Restaurant Las Palmas was just not big enough, especially at weekends when we sometimes had a queue stretching to the end of the block.

I was very proud when the restaurant was considered suitable enough to be recommended by the World Cup committee and earned a place on the World Cup map of Sunderland, and was mentioned in the official guide.

Pepel was very much neglected during those first few months, when we were struggling to get the business on its feet. At first, I tethered her to a tree in the small garden we had in the yard, but I spent so much time untangling her lead that I decided to let her run loose in the yard. This became equally worrying, as she started to climb through the kitchen window to me when she became fed up with playing by herself. Then came the day when she began to venture outside the yard into the lane at the back of the restaurant. She also went along the wall and into the garden of a house occupied by a couple of old people further along the street. The old folk did not mind; in fact they told me that they looked forward to her visits when they would watch her steal their mint out of the garden, or whip the clothes pegs off the line!

Very often we would have people knocking at the door to tell us that we had a monkey on the roof. They looked in amazement when told that we were aware of the fact. Some walked away feeling foolish, others obviously thought we were crackers!

One day I took Pepel to the Argo Frigate, a public house I frequented in town. Their cat had just had a litter of beautiful kittens. They were all black except one, which was tortoise-shell and a male. When the mother was not looking, Pepel picked the odd man out from the litter and started to cuddle it. She loved it from the start, so much so that we had difficulty in taking it away from her. The publicans very kindly promised that I could have the kitten for Pepel when it was old enough to leave the mother.

I did not realise then how grateful I was to be for the kitten when it came to the restaurant to keep Pepel company when I was too busy to do so. I had bought a cage for Pepel, in which she spent many a lonely evening while we were busy in the restaurant. I hated to see her in the cage, just as I had hated it when I put her in the one on the *River Afton*. She looked so lonely and miserable, even though she had a television and an electric fire all to herself. The kitten proved to be the answer to ending these periods of loneliness for Pepel.

# Chapter Fifteen

Timmy, as we called the kitten, arrived on a bitterly cold evening when I had three electric fires on in the room to keep Pepel warm. She was sitting in her cage when I took Timmy up to the flat and her sad little face lit up with joy when she saw him; she could hardly wait for me to put him in the cage with her. The entire staff, which then numbered twelve, came into the room to see Pepel greet her new friend and they were amazed at the instant affection that Pepel and Timmy had for each other.

Pepel pulled Timmy into the cage and cradled him in her arms just like a human being. She then turned him on his back and looked through his fur, inch by inch, making sure that he was clean. He went through this treatment without a murmur, then snuggled up to her as if he thought that Pepel was his mother.

Pepel had not liked her cage any more than I liked to put her in it, but with Timmy to keep her company she actually looked forward to going in it, where she would sit and cuddle him for hours, or they would both sit and watch television in between cuddling sessions. I soon had a monkey who ate meat and a cat who was partial to fruit. They shared whatever was put in the cage. I felt more contented to leave Pepel during our busy periods, knowing that she had a good friend for company.

Timmy soon found that Pepel intended to take her mothering duties seriously. When he became old enough to wander in the street he soon found himself a few girl friends. Sometimes they used to call for him and sit on the wall to wait for him. When he went on his courting expeditions Timmy would sometimes be gone for two or three days at a time. On his return, usually in a hungry and bedraggled state, he had to face his 'mother', who would cuff him on the ears and chatter away to him as if to say, 'You naughty boy, where do you think you have been?' Pepel would then set about his fur and go through it from head to tail, just in case he had picked something up during his travels. Timmy would sometimes protest at this ear cuffing and cleansing treatment by showing Pepel his claws; this however did not deter his zealous foster mother, who would give him another clip on the ear for his audacity. Timmy eventually took it in his stride and knew what to expect when he came back from his girl friends.

I soon realised that Timmy was not going to be much use as a companion for Pepel if he was allowed to disappear for days on end, so I decided, after much deliberation, to have him doctored. He was still under the anaesthetic when the veterinary surgeon brought him back from the clinic. We put him in the bathroom where he was able to sleep it off, to come round in a very dazed condition a few hours later. Poor Timmy could not understand what had happened to him and Pepel was curious to know where he had been and why she was not allowed to go to her 'baby'. When Timmy was fully recovered we let her inspect his operation and give him a cuddle.

I had been warned that Timmy would still 'go out' for a few days after the operation, until he realised that he was unable to satisfy his girl friends. Timmy did wander off for about a week, during which time he just could not understand why his former tabby admirers ran away from him, as if he were some kind of freak. None of the cats we had seen him with would have anything to do with poor Timmy, making me feel very guilty, as well as sorry for him.

Timmy grew and grew during the next few months and his coat became very beautiful. This was because he was content to sit on the wall with Pepel when the weather was nice, or curl up and go to sleep with her indoors. He never ventured further than our own backyard except for the times he accompanied Pepel when she was taken for a walk by me.

Pepel made Timmy famous when they were both photographed while sitting on top of a barrel organ in Union Street, Sunderland. They were 'borrowed' by a friend of mine, a member of the Royal Air Force Association, who thought they would make an unusual attraction during Wings for Victory Week. They certainly did cause some excitement in town that day, when crowds of people blocked Union Street from end to end, all trying to get near enough to the barrel organ to be able to see a monkey with its arm around a cat, a most unusual sight indeed. Apart from filling the collecting tins with much-needed money, the kindly North-Easterners also bought fruit and nuts from stalls nearby to feed Pepel with. They all thought it was very funny when Timmy ignored the kipper someone gave him whilst giving Pepel a hand with her fruit. The picture of this incident, taken by a local newspaper photographer, is one of my prized possessions.

The photograph and article which appeared in the newspaper brought the restaurant extra customers from as far away as Newcastle, Durham, Chester-le-Street and Washington. We were also inundated by numerous telephone calls from old ladies who wanted to give my pets a 'good home' and by children who came to see Pepel and Timmy, some of whom also wanted to see the barrel organ!

Ray's sister, Cathline, made it possible for me to discard Pepel's cage by coming to monkey-sit for me in the evenings. This made me happy and allowed Pepel to enjoy more freedom. Pepel used to sit on Cathline's lap to watch television while Timmy curled up at her feet. Sometimes Pepel would get tired of watching television and would tantalise Timmy until he agreed to play with her. Then they annoyed their chaperone by jumping all over her, usually while she was watching a good programme.

When it became possible for me to employ an extra man in the kitchen I could get out more often to visit local pubs and clubs, making friends and advertising the Las Palmas. I took Pepel on these occasions and she soon became well known in the nightclub scene. At the 'La Strada', a well-known Sunderland cabaret club, she met many of the great stars of today. I was careful not to take her during a cabaret performance in case she stole attention from the artiste.

'Parrot Face' Davis, the well-known comedian, made use of Pepel's presence at the club one night when he spotted her drinking out of her special glass at the bar. He interrupted his act to announce that one of the 'Monkees' had made a surprise visit to the 'La Strada'. This raised a laugh from the audience. After his act, Mr Davis came over for an introduction to Pepel, who was keen to know what the comedian had under his funny hat. Mr Davis always brings back happy memories of the 'La Strada' to me when I listen to his radio programme.

During the summer I used to take Pepel to our favourite spot in Mowbray Park. We always called in at the Fish Bowl on our way there to see Sammy and to give him a few nuts and raisins. Although being in a cage made it difficult for poor Sammy, he did his best to make advances to Pepel, who always remained aloof, preferring to look at the gaily coloured birds in their cages on the wall. Sammy was a very beautiful monkey; I often wished that I could have him, even if only to get him out of his cage, away from the poking fingers he had to contend with day after day.

Another pet shop in the town had a very beautiful male monkey. He also became very amorous towards Pepel when I took her to visit him. Pepel, however, was only interested in his toys lying at the bottom of his cage. He became enraged when Pepel ignored him while trying to steal one of his toys. I could not blame the poor chap, for his playthings were the only comforts he had in a life spent behind bars. Pepel, of course, could never understand his anger or his overtures, but this is because she thinks she is human and prefers human love to that of her own kind.

She fell in love with our young second chef, Brian Ambler, who started with us towards the end of the first year in business. He spent a lot of his spare time with Pepel and was soon able to do what he liked with her. When Brian came to live at the flat with us he took his turn to monkey-sit when off duty. This pleased Pepel because he 'played rough' with her and allowed her to jump all over him. Brian would also take Pepel for a walk when he visited his mother at Pennywell. The first time he did this he had almost got there when Pepel broke loose from him and came running back home. Someone knocked at the door to tell me that Pepel was on the roof, just as Brian arrived back, puffing and panting, to tell me what had happened. Although I was shocked to know that Pepel had run loose from Pennywell to Pallion, I was secretly pleased to know that she knew where she lived.

As the months went by Pepel looked upon Brian almost as her brother. He was able to handle her with as much confidence as myself and even to this day Pepel will look round expectantly when his name is mentioned.

When the business prospered we started to give free meals to old age pensioners. Apart from the big parties we gave them at Christmas, we also entertained four different old folk to dinner every Monday evening. They came from the various 'Over 60s' clubs in the town and soon looked forward to their evening at the 'Las Palmas'. Pepel and Timmy soon became a favourite topic of conversation on these occasions, when I would take them both into the restaurant to meet the aged guests. Pepel sometimes sat at the table to receive tit-bits from them. I received lots of invitations to take Pepel and Timmy to their clubs but I never got around to doing so.

Sometimes I took Pepel for a walk to the Alexandra Bridge, where we would both look at the ships berthed nearby, some of them new ships waiting to be fitted out. I always had a feeling of nostalgia on these visits to my former life. It was then almost two years since I gave up the sea to take on the responsibilities of a businessman. The life of a chief steward was cushy to what I had taken on and I began to envy the men who would soon be taking these new giants away to sea, probably to lands I had visited but would never see again. Pepel seemed to know the yearning I had in my heart and understood the reason for our frequent visits to the shipping scene. She would look excitedly at the vessels moored on each side of the Wear, then turn to me as if to say, 'Well, which one is ours, and when are we sailing?'

At that time I never really thought that Pepel and I would go back to sea again. The sea was part of the past and no matter how much I longed to taste once again the life I craved for as a boy, and loved so much as a man, I knew that I could never forsake the new life I had chosen and the business I had worked so hard to develop and maintain. I had made so many friends in Sunderland that I really seemed indispensable at the 'Restaurant Las Palmas'. Customers complained when 'Tiny' was not there to open the door for them, or to cook some special dish at a minute's notice.

How wrong I was to suppose that my life at sea was finished became evident a few months later, when the whole world seemed to crash around me. I was not to know, as Pepel and I reminisced on the banks of the River Wear, that we were indeed to be claimed once again by the sea and, for a while, would be leaving the Sunderland scene. Pepel did not know then that she would be the centre of attraction in many different lands before her fourth birthday.

# Chapter Sixteen

Although the restaurant had steadily prospered for almost two years and was reckoned to be the most popular eating house in the North-East by our regular customers, the business took a turn for the worse in October 1966. We noticed a steady decline in the luncheons we served. Workers who had been coming to the restaurant ever since it was opened only came two or three times a week, then some of them stopped coming altogether. The lunchtime takings dropped by fifty per cent in less than a month. Our evening trade continued to be profitable until the end of November, when the Labour Government's now infamous 'Credit Squeeze' took its effect, not only on many businesses, but on everyone in the country.

Losing the workers had worried me because we had always tried to give them a good meal at a very moderate price. I thought that maybe we had offended our customers in some way. There are many ways in which customers can be lost in the catering trade apart from bad food and poor service. I once lost two very good customers by just refusing to accept a drink one of them offered to me. They stayed away for weeks because they thought I was 'being funny'.

The 'Las Palmas' was not to blame, however, for I took the trouble of finding out just why our trade had slackened to an alarming degree. I went to see several of the workers and asked them point blank why they no longer came to the 'Las Palmas'. All of them gave me the same answer: that they could no longer afford to have a hot meal during the day. Thousands of workers had been put on short time because of the financial squeeze. The same thing applied to our evening trade. People who formerly came out two or three evenings a week and who normally came to the restaurant for a meal were, they told me, feeling the pinch like everyone else.

Mr Harold Wilson's big economy drive and the various extra taxes and levies he made the catering industry bear were responsible for the ultimate failure of the business. I too was to blame because of my 'softness' as a businessman. In pandering too much to the customer I had put too much back into the business and not enough in the bank. Like most businesses, the 'Las Palmas' had been operated on credit accounts. To enable us to cater for increased trade, up to the big squeeze, we had ploughed back all the profits in the purchase of new equipment, furniture, necessary alterations to the premises and advertising, etc. In two years we had ploughed back almost eleven thousand pounds into a business we started with practically nothing! It was through this that we had nothing in reserve for a rainy day.

Despite the fact that we were almost broke, we decided to carry out our promise to give a hundred old folk a free Christmas dinner that year. Although we had to cancel the weekly free meals for our old-aged friends in the New Year,

I felt that we could not disappoint them at Christmas. They were invited for December 28th when most of our bookings had been catered for. Transport was provided by the local Round Table movement, who had arranged to bring the guests to the restaurant and then take them home again after the party.

I had a lump in my throat when all my old friends sang, 'For he's a jolly good fellow', after the biggest tuck-in they had probably had in twelve months. It always amazed me to see just how much an old person could eat on these occasions. The staff, who gave their time at the party for nothing, were lucky if anything was left for them.

Pepel and Timmy were there as usual to accept tit-bits from the delighted old people, who had pushed back their chairs to shuffle round to the music of the juke box. Others gave a hand to wash up and tidy up the kitchen, while the staff and some of their friends started their bottle party upstairs. Although I managed to put on a show of being happy, they all knew that I was very worried and did not blame me when I too 'went on the bottle' after saying goodbye to the old folk.

The creditors could not have chosen a more difficult time than January and February to press for their money. Suppliers who had given me extended credit and had never before pestered me for payment now worried me to death. It was a case of passing the can. Their creditors were also pressing them for payment. The last weeks were spent in running round from one office to another. From the bank manager to the solicitor, then on to the accountant, trying to chase him into producing the previous year's trading accounts in order to try to raise a loan to carry on with.

All my efforts were futile. The money the bank lent me was sufficient to pay only a few creditors. The bank could lend me no more because of the squeeze. In desperation I wrote to the Prime Minister, only to receive two pages of typewritten codswallop from one of his junior secretaries. People who offered to help me only made me feel worse. One young policeman offered me his savings of ninety pounds. Another man who was on the dole offered to work for me without pay until the business improved. Someone actually sent me a five pound note in an envelope anonymously. I suspected that it was from a pensioner. More than one friend offered to put me up if ever I needed somewhere to stay, but I knew that if and when the end did come I would never be able to stay in Sunderland.

I kept a full staff on as long as possible, hoping for a miracle to happen, but, of course, miracles only happen in fairy stories. It occurred to me that more than a magic wand would be required to pull us out of the mess we were in. Business came almost to a standstill on weekdays, so I decided to close during the day and open the restaurant at night and at weekends. It was with great reluctance that I did this, knowing that the staff would suffer through loss of wages, but something had to be done to cut down expenses.

By the end of February 1967 we were only open at weekends and despite the fact that we had economised in everything except the quality of our cuisine, the takings were not enough to pay for the running costs of the restaurant. Things began to look very black indeed and the word soon spread throughout the district that we were 'on our way out'. There was no consolation in knowing that other businesses in Sunderland were on the brink of disaster also. One other catering concern collapsed before we did and another fell by the wayside just after the 'Las Palmas' served its last meal.

It was with a heavy heart and a touch of irony that I decided to close the restaurant for the last time on April 1st,

1967. On April Fools day! Five months before I had laughed at the offer, made for the business by a Chinese concern, of ten thousand pounds and a salary of thirty pounds a week if I stay on as chef/manager. At that time the business was still booming. Our popularity was at its peak and we were rushed off our feet taking money I had never dreamed of taking when we first opened. I had obtained a licence to serve drinks and, despite some considerable opposition from local people, had managed to get this extended, enabling us to serve drinks until midnight. We were then serving meals until sometimes three o'clock in the morning. Regular customers began to complain because they could not get into the place. Then, twenty thousand would not have bought the business, but a similar offer made on April Fools day would have been the miracle I was looking for!

Given another six months' trading, we would have been out of rough water. Our debts would have been paid and that alone would have been reward enough for me. As it was, we had indeed been fools to work so hard only to have everything snatched away when success was so near.

The last two weeks in Sunderland were like a prolonged nightmare to me. I had put the affairs of the restaurant in the hands of my solicitor and all that remained for me to do was to hang around until the business was disposed of, like a beautiful ship being sent to the scrapyard!

Although I realised it would do me no good, I started to drink heavily, sleeping as much as possible and trying to forget that I was alive and had to face up to my responsibilities. Do not get me wrong, I never felt sorry for myself, only for other people who I felt I had let down. Only a poor workman blames his tools when something goes wrong with the job, but bitterness had crept into my mind knowing that I had failed, not through lack of hard work on my part, but through circumstances forced upon me by a Government which rates the catering trade as 'secondary'.

All I wanted to do was to get away from the scene of my failure and Sunderland, where I could no longer hold my head up high. I had worked with practically no rest for over two years and my body ached with tiredness. My worries had taken a different shape, from worrying about the business to wondering what the future had in store for me, and of course, Pepel, who seemed to realise that something drastic had been going on and anticipated a move. She watched every move I made as if she thought I would be tempted to go without her.

Someone told me that it was Pepel who had brought me bad luck and would continue doing so unless I 'got rid of her'. Although like most seamen I am a superstitious person, the suggestion that Pepel is a bad luck omen is unthinkable, although it has been suggested by more than one person since then. Pepel need never worry about her future as long as I am alive. She may never know it, but it was Pepel who kept me sane during those difficult days. I have never and shall never have a friend like my little monkey, who has shared the rough days in my life, as well as enjoying the smooth ones. I am confident that one day Pepel will be the one to bring good luck to both of us. Whatever we encounter on the other side of the horizon will be decided by Him who created all of us.

# Chapter Seventeen

Deciding where to go did not take long. I knew that Pepel and I would be made welcome by Ruby and 'Striker' Wells, old friends of mine at Great Yarmouth, Norfolk. They have treated me like one of the family for sixteen years and I knew that, rich or poor, there would always be a bed and a Norfolk dumpling for me there.

Another reason for deciding to go to Yarmouth was that I might be able to get suitable employment during the holiday season, or even a ship. But first I wanted a much-needed rest, weeks and weeks of uninterrupted peace, to try to forget what had happened and make plans for the future.

Brian Ambler decided to go to Yarmouth with me, as there was no hope of getting employment in Sunderland. I was glad he made this decision because he proved to be of great assistance to me where Pepel was concerned. Ray drove us down to Yarmouth in an old car he had bought and, up till then, had managed not to smash up. Although he was a good driver, I kept my fingers crossed and hoped that nothing would happen on this journey.

Timmy slept on a rug underneath the back seat for most of the way, only waking up when we stopped for refreshment, when he lapped a saucer of milk. Pepel, who loves a car ride, never moved from my knee except when we pulled into a layby, where she hopped out of the car to stretch her legs and do the necessary. She is never any trouble when travelling, unlike most children, who want to crawl all over the car all the time, or ask for this or that. People often remark how well-behaved Pepel is, no doubt thinking that as she is a monkey she ought to be jumping about all the time.

It was a glorious day so we did not mind the long distance we had to travel. The journey took nine hours, but this included the frequent stops we made for food and to let Pepel have a run in the fields. I wondered what she would think when she realised she had to sleep in a strange bed that night. I also wondered if she would be welcomed as much as myself when we arrived at Great Yarmouth. I need not have been so apprehensive because, although they were all a bit dubious at first, the Wells family soon took Pepel to their hearts.

I had not warned Ruby of our impending arrival. I wanted to surprise her, knowing that she is not the type of woman who goes into hysterics when caught paperhanging or something by uninvited guests. We were received by the Wells in their usual warm and friendly manner. The mere mention of my longing for a Norfolk dumpling, drowned in onion gravy, sent Ruby into the kitchen to make one for me. Pepel developed a palate for this Norfolk speciality too and gorges herself with handfuls of dumpling whenever she gets the chance.

Ray stayed for a couple of days before going back north

I was sorry to see him go, after trying to persuade him to stay on in Yarmouth, knowing that he would stand a better chance of getting a job there than in Sunderland. Although he had sometimes been very reckless, he had lost just as much as I had when the business collapsed. I hoped that he would be able to settle down and carve a better life for himself and when we met again we would both be in better circumstances.

Timmy settled himself into his new home, much to the disgust of Mandy, Ruby's pet dog. I remembered the day that Mandy became the Wells boys' pet, when she was only a pup. Looking at Mandy reminded me how quick the years had gone by since she used to go with Sidney and Jimmy Wells to watch their father, Striker, eel fishing in the River Yare. Now, Mandy had become too old to move further than the yard, which, naturally enough, she guarded jealously against this new interloper who had invaded her territory. Funnily enough, Mandy did not mind Pepel so much and Pepel would even let Mandy steal her sweets, to which she helped herself daily from Ruby's handbag.

The first day we were at Yarmouth Brian and I took Pepel along the banks of the Yare, where we sat in the sun watching the boats arriving at the Yachting Station from the Broads. Pepel took advantage of her freedom and scampered about in the long grass, returning now and then to see if Brian and I were still where she had left us. Subsequent days were spent in the same way, or on the promenade, where Pepel soon became an attraction. There were plenty of monkeys at Yarmouth, owned by the photographers who try to pester people to have their pictures taken with a monkey sitting on their shoulders. None were quite as beautiful or as interesting as Pepel, who used to steal their limelight. One day, when a crowd had collected to look at her, a police car came screaming to a halt on the promenade and several policemen came over to see what was happening. They thought I was another photographer until I assured them that Pepel was just my pet and, like the crowd around us, we were on holiday. Apparently there was a big purge of the promenade 'monkey photographers' that season.

I soon stopped taking Pepel to the sea front because people who wanted to admire her made relaxation impossible. No matter where I went, or where I sat, a crowd soon appeared as if by magic. It was an education to watch the different expressions on people's faces as Pepel and I walked by. Some would look in amazement at the monkey with a long tail. Others did not even notice her. Some of the people who saw her had to look again to make sure they were seeing right. Most people looked at Pepel with obvious kindness and amusement, while a few shuddered as if to say 'fancy walking around with that thing!' I felt hurt and angry for Pepel when we met such people.

Two weeks had gone by before I received a letter from the Booth Line of Liverpool, which had been forwarded from Sunderland. I had applied for a job with them before I left there but had not expected them to reply after such a long time. They asked me to phone their Liverpool office about a position they might have to offer me as chief steward. When I contacted the superintendent he asked me to go to Liverpool the next day for an interview. I walked on air when I left the telephone box and hoped that I would have some luck in Liverpool.

Pepel and I caught the early train the next day. When we changed trains in London I kept thinking that I had lost something until the reason dawned on me. It was the first time in years that I had gone anywhere without luggage! I saved quite a bit in tips that trip.

I left Pepel with some friends in Dingle while I went for

the interview. It was a nice day so I thought she would be all right in their yard, fastened so that she could sit on the wall for an hour or so until I returned. My friends, who I had not seen in years, were delighted to see me, and promised to look after Pepel. They also gave orders that I must stay the night whatever the outcome of the interview.

It came as a surprise when I succeeded in getting the job. I had been away from the sea for three years, but the superintendent was satisfied that my qualifications were right. He asked me to fly out to New York the following Saturday to join a ship there which traded on the American coast. In the meantime I could stay on board one of the Company's vessels berthed in one of the Liverpool docks and have a medical examination by the Shipping Federation doctor. There would also be my passport and air passage to arrange during the following week.

A great weight had been lifted from my shoulders when I walked out of the office. Although I had been thankful for the rest since leaving Sunderland, the thought was ever present in my mind that Pepel and I could not live on fresh air alone and I needed to get a job sooner or later. Getting back to work would take my mind off the loss of the 'Las Palmas'.

The sudden realisation that I had completely forgotten to mention Pepel to my new employers brought me to a standstill at the end of the street. I began to panic, fearing that they would object to her going to my new ship with me. I knew that her entry into the States would be no problem, as her U.S. health certificate covered her in that respect. Apart from a possible objection from the Company, there was also her air fare to consider. I did not know how much it would cost to take her to New York, but I knew that it would be more than I could afford. Although I would be reimbursed by the Booth Line, the return fare to

Liverpool had made a hole in the little money I had.

I thought, well, the best thing to do is to go back to the superintendent and explain the position to him, then he might help in some way. He did. He informed me that under no circumstances were pets of any description allowed on board the Company's vessels! Apparently there had been a lot of trouble caused when the crew of one ship had brought about sixty South American bushy-tailed monkeys on board, intending to sell them in the States. It seemed that these unfortunate monkeys had run riot and when the vessel docked the fire brigade had to be called in to capture them. This had interested the television people, the police and the Ministry of Agriculture. It also infuriated American animal lovers and the incident, amusing as it was in some respect, had never been forgotten by the Company.

If I wanted the job I had to leave Pepel behind. My mind was in a whirl as I sat facing the superintendent, who, while being firm, was also trying to be as kind as possible, knowing that I had a problem. I asked him if I could think it over during the weekend, even though I knew then that Pepel would have to come first. Their callous treatment of the bushy-tailed monkeys was not the only reason why I cursed the seamen responsible for their fate as I walked miserably away to collect my own little friend.

I stayed that night with my friends, Mr and Mrs Gavin, whose son John had been a catering boy with me some years ago. He had married and had had a child since I last saw him, another reminder that the years were flying by and I was not getting any younger! We all went out for a drink and talked about old times until it was time to go to bed, where I went to sleep still wrestling with the problem of Pepel, who snuggled up to me oblivious of what was happening to us.

Pepel and I sat in the station buffet the next morning,

waiting for the train back to Yarmouth. I was still wrestling with my conscience and torn between my love for Pepel and the fact that neither of us could live unless I got a job.

I forget how many cups of tea I drank while I tried to sort things out. Pepel was eating wafer biscuits and watching people coming and going, mostly holiday makers on their way to the coast. Just then something happened to resolve my problem beyond any doubt. An old couple sat at my table who were loaded down with luggage and a pet poodle. Getting into conversation with them was not difficult because they had purposely come to my table to see Pepel at close quarters. When I asked them where they intended taking their holiday they replied that, as they were not able to take their pet into the hotel they had booked, they were returning home. Their holiday, they said, 'would not be a holiday without Tinker'. I did not tell them, but it was they who finally made up my mind about Pepel. I realised that, like their holiday, my life would be no life without her. I also realised that the same problem would arise again, probably many times in the future, but under no circumstances could I forsake Pepel, who clutched my little finger, knowingly, as we boarded the train to Yarmouth.

# Chapter Eighteen

It made me happy when Ruby said that I did right 'to think of Pepel and come home'. Good, warm-hearted Ruby! What a pity there are not more people like her. Just as there will always be a place in their home for me, there will always be room in my heart for Ruby and Striker Wells.

Brian got himself a job as Second Chef at a local restaurant, where I also later obtained a position as Manager. In the meantime I spent my time in writing letters to various companies hoping to get away to sea as soon as possible. I also took Pepel along the riverside to see the ships and supply vessels belonging to the oil companies engaged in the search for gas in the North Sea. I could have gone on one of the oil rigs as chef/manager. The job was offered to me on one of our visits to the riverside. But again, there was no room for Pepel. No animals allowed! I began to realise that it is a load of rubbish about the English being great animal lovers!

E

One day, as Pepel and I were watching the loading of horses and cattle aboard a Continental ship, one of the dockers, who is also a freelance photographer, asked if he might take a picture of Pepel. After I told him to go ahead he took several of her, then invited me to take her to his local pub that night. That is how Pepel came to drink a toast to Sir Francis Chichester, the famous lone voyager. It seemed appropriate for her to do so, being an ex-seagoer herself.

Photographs taken of Pepel in The Ship Inn enjoying a bottle of beer had in the background a large poster which said, 'Drink a Toast to Sir Francis'. Pepel did just that, and got herself into an article written about her in the local newspaper entitled 'Monkey-Business at The Ship'. Her photo also appeared in a well-known brewers' magazine.

Working as manager at the local restaurant, where Brian was chef, kept me going for several months. I thought, however, how ironical it was that I should be there working for someone else in almost the same capacity as I had been in my own place a few months before. The thing that sickened me most was that I was not allowed to give the holiday makers the same quality meals and moderate prices they would have enjoyed at my own restaurant. The way in which working people on holiday are robbed – yes, robbed – of their hard-earned money is scandalous. It is no wonder to me that people would rather spend one week abroad than two weeks at an English resort!

I became so depressed in the job at Yarmouth that it was like a tonic to see Captain Reg Watson standing on the doorstep one day. I have known him for seventeen years and sailed on vessels under his command for a total of ten years.

Captain Watson had received a letter I wrote to him some weeks before, but had been on the Continent, where he had been unable to help me in my search for a ship. He had come to let me know that a relief steward was required on the m.v. *Alacrity*, berthed in Yarmouth at the time and waiting to sail on the tide. He said that the job was mine if I could join the vessel within the hour. I asked him if I could take Pepel on board with me and almost jumped for joy when he said that it would be O.K.

It was the first time I had ever joined a ship 'dirty'. Being caught on the hop with very little notice, I had to cram dirty clothes in my cases instead of the usual freshly laundered gear. Still, there would be plenty of time to dhobi it on board. I was more concerned about getting to the *Alacrity* in case she sailed without me.

I nearly fainted when, on arriving at Fellows Shipyard, where the ship should have been, I discovered that she had already sailed! I stood there with Pepel, who was wrapped in a woollen scarf to keep the cold out, my suitcases on the quayside and the taxi gone, wondering what to do. Just then the ship's agent arrived and explained that the *Alacrity* had had to sail but was anchored in the Roads, waiting for me to be taken out in the pilot cutter. Although the weather was bitterly cold, I was wet through with nervous sweat and was grateful for the warmth of the little boat which took me out to my new ship and, I hoped, to a more hopeful future.

Pepel quivered with excitement as I scrambled up the Jacob's ladder on to the wet, slippery deck of the year-old tanker. She soon made herself at home in the small but compact cabin which was situated aft, just below the modern well-equipped galley, where I found an ordinary seaman already cooking the evening meal in the absence of the cook.

The master, Captain Leslie Bowler, welcomed me aboard later that evening with a glass of champagne. He was much amused to see Pepel also enjoy a drop of bubbly and was amazed at her strength as she drank out of an almost empty bottle. Pepel can lift most bottles, regardless of size, if she fancies their contents.

Captain Bowler would have detained me in his cabin all night in order to watch Pepel and her antics, but with all the excitement and rushing about I had experienced that day I was anxious to turn in for a good night's rest, ready for work the next morning. Taking leave of my host, I took Pepel below to my cabin, where I later went to sleep thinking about Timmy, whom I had had to leave behind at Yarmouth. I hoped he would not be lonely without Pepel, who was already sound asleep beside me.

As the *Alacrity* is a new ship designed to carry less crew than is normally carried on a ship of her tonnage, my department consisted of just myself. I was chief steward, cook and galley boy, making what is known as 'cook steward'. I did not mind this drop in rank, in fact I felt much happier because I love to cook and this job would serve to keep my hand in. I had no catering problems as the ship was 'company fed', meaning that all food supplies were purchased by me through a chandler who was paid by the company. This made the job less worrying than on the vessels where the crew find their own food. I had only ten men to cater for, so it was rather like looking after a big family.

We loaded a cargo in the Thames for Antwerp. While we were waiting for the cargo to be put aboard, the Captain invited Pepel and me to his home in London, where he promised to show me around 'Jack the Ripper Country'. Seeing the London shop windows dressed up with fairy lights and synthetic snow reminded me that it was almost Christmas – December 16th to be exact. I thought about other years I had spent Christmas at sea and wondered where December 25th, 1967, would find me.

Fate brought Captain Watson back into my life when the *Alacrity* berthed in Antwerp on Christmas morning. His ship, the m.v. *Arduity*, was berthed on the other side of the dock, where she had very conveniently broken down the previous day, giving everyone on board Christmas in port. The *Arduity* was not company fed so I was able to repay some of his kindness by sending Captain Watson and his crew some extra Christmas fare, which we had in abundance. Captain Bowler delivered it, as I was too busy preparing and cooking our own turkey and 'duff'.

Although from all accounts they had a hectic time on board the *Arduity*, Christmas came and went very quietly on board the *Alacrity*. The ship pumped her cargo ashore all day, making it a working day for almost everyone on board. After the dinner was taken care of I prepared a buffet tea in the officers' saloon so that everyone could help themselves when they felt like it, although after the large helpings of dinner everyone had consumed I did not expect much enthusiasm for tea.

It was Pepel's fourth Christmas and I was sorry that it had to be a giftless one. I had been unable to buy her anything, but I made sure she had her share of the extra fruit and nuts, etc., available on board. She was quite content to sit with me in my cabin and listen to the radio. People ashore take Christmas radio programmes for granted, but the seaman takes more interest in them because, although he may not be a religious man, the services and especially the carols are a means of bringing home a little nearer.

We were at sea and bound for the Baltic as the year 1967 drew to a close. The temperature was well below zero and it became colder as the *Alacrity* ploughed her way into the Skagerrak and rounded the northern tip of Denmark.

Although we then had land on both sides, Sweden on one and Denmark on the other, it became colder, so cold in fact that the condensation on the inside of the portholes turned into ice and formed little icicles as the moisture dropped from the port boxes.

Until then I had given Pepel her daily exercise on deck, but as the cold became severe I had to keep her in the cabin most of the time. Pepel did not like that one bit and protested by stripping the clothes from my bunk each time I left her alone. There was nothing else I could do. The decks were freezing up and even the handles on the inside of the doors leading on deck were so cold that they burned when touched with ungloved hands! The next few weeks proved very difficult as far as poor Pepel was concerned. They were very distracting weeks for me too, when I tried to do my work and worry about Pepel's welfare at the same time.

I was awakened one night as we were nearing Stockholm, by a crashing and bumping noise which seemed to be on the outside of the bulkhead in my cabin. Pepel heard the noise too and we both lay listening to this strange slithering, bumping and crashing until I decided to leave the warmth of the blankets to investigate. On going on deck I discovered that the noise was made by huge blocks of ice as the ship bumped and ground her way through them. This, I found, was nothing compared to what we had to chop through later. I was not the only one to be grateful on our arrival at Stockholm, where those who were not on cargo watch or other duties were able to sleep without the serenade of disturbed angry ice.

I was sorry that it had to be Stockholm where, for the first time in her life, Pepel was snubbed. I had been looking forward to our visit there because of hearing so much about the place from a group of students I befriended in Sunderland. They came from Stockholm and used to eat at my restaurant. Whenever they were broke (which was more often than not), I gave them a meal at cost price and sometimes for gratis. They were always a polite, jolly crowd of youngsters who fired my imagination about life in Sweden, particularly Stockholm.

I would never have taken Pepel ashore there but for the fact that, above all things, I was out of cigarette papers and, as I only enjoy cigarettes rolled by myself I would either have to get some papers or go without a smoke for the rest of the trip. The Agent told me that there was a kiosk just outside the dock gates, but as this was almost two miles to walk in the snow I ordered a taxi. The taxi driver did not understand a word of English; consequently, when I asked him to stop at the kiosk he drove straight on into Stockholm and stopped outside an expensive looking hotel. On second thoughts, the driver probably did understand English! We seamen will never learn!

Walking into the foyer I asked the girl at the reception desk where the bar was situated. She directed me upstairs as she looked open-mouthed at Pepel, who was wearing a red jumper and Captain Bowler's white woollen scarf. I had to pass through part of the dining room, amid stony glances from some of the patrons, to get to the cocktail bar, which was empty except for the girl behind the bar. Her surprise at seeing a red-jumpered, white-scarved monkey did not deter the lovely Swedish girl from serving me with the most expensive type when I asked for a glass of beer. Instead of giving me a glass of the cheaper beer she reached into the bar's modern cooling unit for a gold-topped fancy-labelled bottle, which cost me the equivalent of six shillings in kroner!

Before I had finished my second bottle, a man wearing evening clothes approached me, talking excitedly in Swedish. Although I could not understand what he was raving about I guessed it had something to do with Pepel, who began to pull faces at the irate gentleman. It seemed that she also knew the reason for his excitement.

Giving him a bewildered stare I told the gentleman, who turned out to be the assistant manager, that I did not understand the Swedish language, whereupon he replied in perfect English, 'I am very sorry, sir, but the management do not allow animals in the hotel, so will you kindly leave at once?' I did not mind being asked to leave, but I did resent the way he looked at Pepel, as though she was some kind of vermin who might spread a terrible disease through his precious hotel! I told him that as he charged too much for the beer he served I was about to leave anyway, but I took my time over the beer I had already paid for. The embarrassed girl behind the bar apologised to me before I left and said, 'Your little monkey is beautiful and I like her'.

I made sure the driver stopped at the kiosk on our way back to the ship, where I nearly collapsed with shock on being charged six shillings and sixpence each for the books of cigarette papers I bought. I learned later that the biggest racket being operated on ships going to Sweden was the smuggling of 'Red Rizla' cigarette papers. Costing threepence halfpenny in England they can be sold in Sweden for three shillings a packet, making a handsome profit, unless the smuggler is caught, when he is fined about a shilling for each paper. As there are sixty leaves in a book the fine would be very shattering to the man caught trying to smuggle a whole carton of a hundred!

Another thing that shocked me in Sweden was the pornographic literature being sold like comics to children. A small crowd of kiddies were at the kiosk when the taxi stopped there. They were buying these lurid magazines, the contents of which left nothing to the imagination, displayed in the window for all to see. I mentioned this to the agent when I arrived back on board and he shrugged his shoulders, saying that the children must have been sent to buy them for their parents. The same magazines apparently fetched a good price in the Kiel Canal; needless to say, as we were going through the canal homeward bound, oil was not the only cargo being carried!

Proceeding to Helsinki we loaded a valuable cargo of alcohol destined for Felixstowe. Navigating the ship must have been very difficult because we had to break our way through ice three feet thick on our way to the bleak Finnish port. As the bows of the *Alacrity* broke a passage through the ice, the broken chunks came sliding to the stern, knitting themselves together again as the propeller pushed them into our wake to make what looked like crazy paving across the lake of ice.

I managed to scrounge a Russian-type hat from one of the installation men who came to admire Pepel at Helsinki. I started to wear it even in bed to keep my ears warm. How Pepel survived that trip I shall never know. The ship was like a huge refrigerator. All the rails were covered with thick ice, so were the masts and everything on deck. In fact, we were very lucky to be able to get away. There was some trouble in the loading of the cargo and the Captain was told that if he did not get away by the next morning, the ship would be stuck there for the rest of the winter.

Pepel looked miserable with the cold. Apart from making her stay in the cabin, where she sat on top of the radiator most of the time, there was nothing else I could do to keep her warm. I was very glad when the Captain decided to leave part of the cargo and get out while the going was good. We followed the special ice-breaking vessel out of the frozen lake, where Pepel, looking out of the porthole, saw something to excite her in the distance long before any of us realised what it was. The small dark speck on the horizon turned out to be a man fishing through a hole in the thick ice. We actually passed within a few yards of him as he pulled a big fish out of the hole and held it up for us to see. Pepel was so excited that she fell off the port box, chattering away at this unusual sight.

We were all glad when the temperature began to rise, but we were at the Brunsbottel end of the Kiel Canal before the ice on the decks began to melt. We still had ice on the ship when we arrived at Felixstowe, where I was relieved to take a month's leave. I had not expected to get a relief so soon, but I was grateful for the money my trip to the Baltic had brought me. Captain Bowler gave me a beautiful briefcase as a farewell gift. He also gave Pepel a box of liqueur chocolates and his woollen scarf which she had worn all the trip. Pepel seemed to know about the gifts because for the first time since she had known him she allowed the Captain to stroke her without pulling a face at him!

My wallet felt unusually thick in my hip pocket as the agent drove Pepel and me to the station. Yet another stage accomplished, I thought, towards a still unknown future. I knew I had to take care of my 'pay-off' in case I had to wait a long time for another job. The agent told me that the *Alacrity* would be going to Lisbon after repairs to her propeller, which had been damaged by the ice at Helsinki. Just my luck, I thought, to miss such a trip after two months in the Baltic. Pepel too would have enjoyed a trip to the warmth of Lisbon. Ah, well . . . !

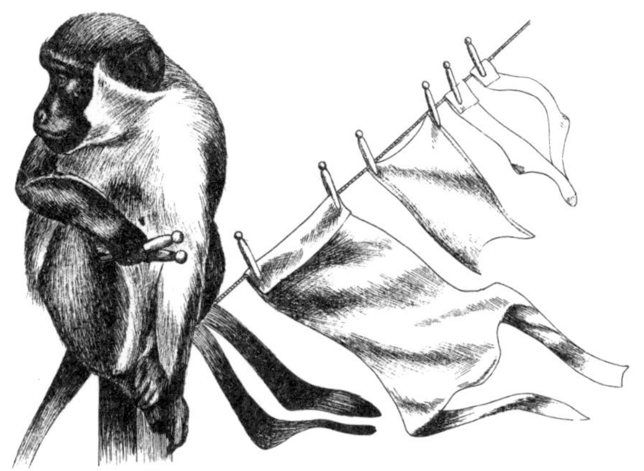

# Chapter Nineteen

I had left a message with Ruby's next door neighbour to ask Brian to meet Pepel and me on arrival at Yarmouth. I was surprised to find that he was not at the station when the train pulled in. Although Ruby's house is not far from the station, I had to take a taxi because I had so much gear to carry, apart from Pepel.

It turned out that Brian had gone home to Sunderland only a few days before, after working as a chef on one of the oil rigs for a month. Apparently the twelve-hour shifts, seven-day-a-week job had been too much for him. I have heard that although the pay is good the work is very hard on the rigs. I suspected that the rough weather had been another factor in Brian's decision to leave. One of the rigs had broken loose only a few days before and knowing the temperament of the North Sea I can imagine what it was like.

Timmy the cat was overjoyed to see his 'mother' return

from the sea. He rubbed himself all over her, purring away as if she had been away for years instead of two months. Pepel soon disposed of the initial welcome to throw him on his back to give him some treatment. She went all through his fur, chattering away to him as if to say, 'Now that I am back you are going to do as you are told, my lad'! Timmy had been well looked after during her absence and had got bigger. He was also becoming an excellent rat-catcher.

A couple of days later Timmy came into the yard with his tail almost hanging off. It looked as if he had been savaged by a dog, or had received a vicious kick from someone. Whatever happened we shall never know, but he was in terrible pain and would not let anyone touch him. After a while I managed to catch him by using Pepel as a decoy. He went to her miaowing miserably, pleading for her help and understanding. Although he struggled violently he kept his claws in and never attempted to scratch me while I inspected his wound. With the help of Sidney Wells I managed to bathe his gashed and broken tail with T.C.P., then bound it up as best I could with a finger bandage. We kept him locked up for a few days until the wound knitted together, but it was quite a while before he was able to lift his tail proudly as he used to do. He spent most of his time with Pepel, sitting on the fence or on top of the coal bunker in the yard.

A more amusing incident occurred one day when Pepel was sitting on the fence in the yard. Someone next door went to use the outside toilet adjacent to the wall separating the two houses and near to the corner of the fence Pepel was sitting on. Pepel became curious and lifted a loose tile on the roof of the toilet, making clacking noises as she lifted the tile higher to get a better look at the embarrassed occupant.

Mark, the little boy who lived next door, loved Pepel and although he could hardly see her unless he climbed the wall, he used to stand in the yard and talk to Pepel for hours. One day when Ruby went to hang some clothes on the line, she found that Pepel had stolen all the clothes pegs, whereupon she gave forth at Pepel, from a safe distance, telling her that she was a naughty girl and would get her bottom smacked. A little voice on the other side of the wall told Ruby that she would get her bottom smacked if she hurt Pepel! It was, of course, Pepel's little friend Mark who was sticking up for her.

February gave way to March and I received news that I might be required to join Captain Watson's ship, the *Arduity*. The news came over the telephone from the Captain's wife Iris, whom I have known as long as the Captain. She said that the *Arduity* might be needing a steward when the vessel arrived at Middlesbrough from the Continent. She would let me know for sure when the Captain telephoned her that night.

I felt very excited and happy at the thought of going back on a vessel under Captain Watson's command. We had been good friends for years and I knew that he would like Pepel. I hoped and prayed that the phone would ring soon and even started to pack in anticipation.

Neither Pepel or I knew then that something would happen in the near future to bring us closer together than ever before. The most dramatic and agonising incident that ever happened in my life occurred on board the m.v. *Arduity*.

# Chapter Twenty

It was the second time I had been steward on board the *Arduity* when I joined her on March 14th, 1968. The first time was about ten years ago, before she was altered and made bigger with extra tanks. She was the pride of the fleet then, but when I saw the state she was in as I joined her in Middlesbrough I could not help feeling dismayed. Still, I thought, the poor old girl has worked hard and, being now over twenty years of age, is feeling the strain.

It was breakfast time when Pepel and I arrived on board. We had travelled since nine o'clock the previous evening from Yarmouth and although Pepel had been able to sleep a little I had been unable to get my head down. I felt very tired as the dock master helped me aboard.

Captain Watson greeted me on the bridge as he gave orders for the ropes and wires to be cast off. It seems that they had been waiting for me to join but had had to move into the lock because of the tide. We steamed out of

Middlesbrough ten minutes later, bound for the Thames.

Another old friend I found on board was the chief officer, Jack Randleson, whom I had sailed with many times before in my early days. When the ship was under way we were able to talk about old times over a steaming mug of tea. Jack's main interest was how I came to have a monkey. Captain Watson's only interest was in a decent meal. Apparently they had been unfortunate in having a bad cook and were all half starved. I promised that when I had refreshed myself with a little nap, I would see what could be rustled up by way of nourishment.

My cabin was on the boat deck aft. A small cabin but very comfortable and private. The galley was situated on the same deck just round the corner from my cabin, so I did not have far to go to work. As on the *Alacrity*, I was cook/steward, but on the *Arduity* I had a catering boy to assist me. He looked after the officers' accommodation and served the meals, leaving me a free hand to do the cooking, order stores and look after the ship's linen, etc. It was really a very comfortable job which I could do with ease.

There was no catering boy on board when I joined, so everyone mucked in with the chores. This included the Captain who volunteered to do the washing up. He was suitably rewarded with his favourite curried dishes later on when I had settled into my new job. It felt good to be on a ship where everyone liked to help each other and my pet would be able to settle down in a nice atmosphere. Pepel has always reacted to atmosphere. She knows instantly if it is friendly or hostile, either in regard to her or to myself.

Pepel began really to enjoy life on the *Arduity*. She took to Captain Watson immediately and he soon became very fond of her. He made a special run for her on the bridge deck, where she spent most of her time during the day when I was busy. She soon ventured into the wheelhouse, where she would sit for hours staring at the horizon as the vessel ploughed towards it. She enjoyed it most when the Captain took a watch, when he would talk to her in between plotting the course or sending out radio messages.

Pepel went to so many places and met so many people during the time we were on the *Arduity* that it would require another book to record everything that happened. So I shall only write about happenings most vivid in my mind, leading up to the day Pepel fell into the tank described in the prologue.

As the weather became warmer Pepel was able to play on deck more often. The run made for her was on the port side of the bridge deck where the 'jolly boat' was kept. Sometimes we would put her on the starboard side, depending on which side was warmest, but she preferred to play in the boat on the port side, where she could also look through Captain Watson's porthole to watch him doing his paper work. The Captain's bunk was opposite this porthole and he very often woke up to find Pepel peering in at him. He did not mind this audacious breach of his privacy and enjoyed hearing her chatter to him.

On looking in the jolly boat one morning I saw a collection of ball-point pens and pen caps lying underneath a coiled rope at the bottom of the boat. When I showed these to Captain Watson he laughed and said, 'They belong to the agents, berthing masters and different officials who coax Pepel as they come aboard'. Apparently she allowed them to coax her while she rifled their pockets. Pepel became known as the 'pen-pinching monkey' to these now wary visitors to the ship. Some of them allowed her to steal their pens just for the fun of it, which was annoying, as I had spent so much time in teaching her not to steal other people's property. Apart from this, Pepel was liable to give them a nip when they playfully tried to retrieve items they had let her take from their pockets, which was not so amusing for them.

Our agent lost his spectacles when he came aboard at Preston. I did not actually see what happened, but the incident was described to me by one of the installation workers who had become very fond of Pepel during our visits there. He told me that the agent had gone on board by using the ladder situated midships. After he had completed his business with the captain, he had come aft, when he spotted Pepel sitting on the rail amusing one or two port workers. Without any warning he approached Pepel and coaxed her on to his shoulder, whereupon she mischievously took his spectacles and jumped to the deck to play with them. Having his glasses snatched naturally frightened him and he became perturbed when Pepel, now in her element, refused to return them. One of the onlooking workers, a very brave man it seems, jumped on to the ship and tried to deprive Pepel of her prize. Rather than let him take the spectacles away from her, she did what all monkeys do, she threw them away. The only trouble was that they went over the side into the dock!

I was furious about the whole incident. In the first place I had put Pepel on the after deck, away from the official access to the ship so that she would be no trouble to people coming on board. The agent went to her at his own initiative and he should have known better than to pet a strange animal without even the permission or protection of the owner. The brave man who tried to get the spectacles back should have known better also. He had disappeared by the time I arrived on the scene, but I would have asked him if he would be brave enough to take a bone from an Alsatian or meat from a lion. The incident would have been passed off as an amusing one if someone had called me to retrieve the spectacles. As it was, the people concerned were very lucky not to be bitten, in which case Pepel would have been branded as a vicious monkey. The golden rule where any monkey is concerned is never take anything from them, even though it may be your own property they have mischievously stolen. Better still, do not attempt to pet a monkey (or any animal) unless invited to do so by the owner. Many people have wanted to cuddle Pepel – indeed, many people have been able to do so – but I always advise everyone to admire her from a distance and go away loving her.

Pepel soon forgot about the incident as she played on a plot of grass close to the ship, where she soon attracted crowds of dockers who came to see 'the monkey who pinched a pair of glasses'. Word had soon got around and it became a joke to everyone, except the unfortunate man who had returned to the office minus his spectacles.

It amazed me to hear that they knew about the incident even before we arrived in the Manchester Ship Canal. If Pepel had laid a golden egg or given birth to the three wise monkeys no one would have heard about it, but everyone at Ellesmere Port seemed to know about her disgracing herself at Preston! The pilot who took us into the canal had a good laugh about it, saying it reminded him of a story he had heard about another monkey.

Apparently on one ship a member of the crew had fetched on board a baby monkey and it became the pet of everyone in the messroom. The monkey grew and grew and, as it got older, it became more mischievous and sometimes vicious. After six months they realised that their pet was a baboon and it became such a nuisance that they decided to put it into a cage. The baboon broke loose one night and smashed everything in sight, including the home-made wooden cage. It went berserk all over the ship and did quite a lot of damage, so much so that the captain ordered the poor baboon to be destroyed.

Although the crew had lost their love for their now dangerous pet, no one had the heart to kill it. After a few days had elapsed and the baboon was still on board, the captain again ordered the crew to get rid of it. They decided that the only way to decide who was to perform this task would be to draw lots. This they did, and it fell to one of the able seamen to carry out the execution.

The next day the able seaman was seen making a raft. It took all day for him to complete, after which he lashed a strong box on to it and secured to it a ten-gallon drum of water. He filled the box with all kinds of food prior to lowering the raft into the sea, then he brought the baboon out of its solitary confinement and put it on the raft. It was too busy eating some of the food to notice that the raft had been cut loose to drift away from the ship, sadly watched by the crew, until it disappeared beyond the horizon.

I thought it was a beautiful but tragic story and had no doubt about it being true. Most seamen have soft hearts underneath their rough exterior and I could well imagine the struggle going on in their minds when the execution of the baboon proved too much for any one of them to carry out. I also understood how they must have felt as the raft, bearing the poor animal, drifted away from them.

We did a couple of trips to Eire during the outbreak of foot and mouth disease in England. The Irish official who came on board at Limerick was most perturbed when he saw Pepel. He said, 'You won't be able to take THAT THING ashore you know'. Calling Pepel 'that thing' got my back up immediately. I said, 'Why not, is there some disease she might catch in Limerick?'

Pepel did in fact go ashore after another official pointed out that monkeys were 'not on the list', whatever that may have been. She was issued with a landing card like everyone else on board and had to walk on the disinfectant-soaked straw, which had been scattered at the foot of the gangway. It happened to be Good Friday and all the pubs were shut – that is, officially they were not open for business – but considering the amount of drunks that were hanging around in their vicinity, at least some of them had the back door open!

I too was looking for a drink; so was Pepel for that matter. When I asked a passer-by if there were any hotels open where I could get a drink, he told me that if I took my turn 'on the Corner' I would get the signal when to go to the back door. I did hang around for a while, more for amusement than the desperate need for a drink. It was just like watching a James Bond film. A proper cloak and dagger Good Friday alcohol service! I thought how funny it was for these people to take so much trouble to conceal the fact that they were drinking 'in secret' when it was so obvious what they were doing. The very smell of beer was all over Limerick!

It was too cold to stand around for long so I made my way back to the ship, where the captain, chief officer, Pepel and I had a little party in my cabin. Pepel did her party trick of opening the cans of beer for us while we talked about old times and gossiped about other people we knew that worked for the same company. Seamen are great people for gossiping. Much worse than the average woman.

We were sailing that night so Captain Watson did not stay long. He wanted to get a couple of hours' sleep before the pilot came on board. Pepel soon gave me the hint that she wanted to do the same by pulling down the covers of my bunk, to lie with her head on the pillow, waiting for me to turn in too. Although I am a light sleeper I did not hear the engines as the *Arduity* slipped away from Limerick to start her journey back to the Thames.

We had no orders when we left Limerick, but Captain Watson brought me the news that we were bound for Sunderland from the Thames. He received the orders on the radio and knew that I should be excited at the prospect of seeing my friends there. I was more than excited – I was thrilled to death! I wrote a quick note to Brian and one to my ex-partner Ray to let them know that I was about to descend upon them. As it is only a thirty-six hour journey by sea from Coryton to Sunderland, I hoped that the G.P.O. would be able to deliver the letters before we arrived so that Ray and Brian would be on the quay to meet me and, of course, Pepel.

It was a pleasant trip up to Sunderland. The weather was perfect for the Easter holiday and I looked forward to a little time off when we arrived. There had been quite a few catering boys joining the ship since I had been on board. Some of them were first trippers and left because they were sea-sick. Others had left because they were downright lazy and did not like the work. The boy who was on board for the Sunderland trip had proved himself to be a good lad and willing to learn. He had only been on board a few weeks but already he could do a little cooking and was improving his knowledge of ship's cookery by experimenting in the galley during his off duty periods.

I was grateful that I had such a good lad to assist me, and could now leave him to do the job. I prepared all the meals the previous evening so that he just had to warm them up during my absence in Sunderland.

I have never seen Pepel so excited as she was when she saw the familiar lighthouse at the mouth of Sunderland's harbour. She had been to Roker many times before, but never in such a state of excited anticipation as when we steamed past the fishermen on the pier and on into the Hendon dock. Her whole body quivered as she sat on the rail while the vessel was being berthed. She kept looking round to me as if to say, 'Well, where are they?' I was thinking the same thing and feeling very disappointed that Ray and Brian were not there to meet us. Still, I thought, maybe they will be down to see us later. Perhaps they were working, or even away on holiday.

They arrived as I was just about to lock the cabin door and take Pepel ashore to find them. Pepel saw them first and nearly went mad! She nearly choked herself with the noises she made as she threw herself first at Brian, then more calmly at Ray, to settle in his arms uttering little croaking sounds, pleading with him to cuddle her.

' Ahyeaalreetlad ? Ahseeyavstilgothamoonki ', was the question and observation made many times to me that night by old friends in the town. It felt good to be among the Geordies again, even though the visit was to be a brief one. We were sailing again in the early hours of the morning.

Pepel was soon surrounded by old admirers. I could tell that she felt at home, listening to the language she was familiar with and helping herself to the froth of many beers. The Geordies like their beer and are particular about the froth, which must stand at least an inch above the rim of the glass, otherwise the beer is not fit to drink!

Pepel had learned a few simple tricks while she had been away from Sunderland. She kept her audience spellbound, bringing her act to a close with her favourite trick of 'cherry stealing'. The barman asks her if she would like a cherry. Taking her cue by sitting upright and holding out a hand very daintily, the barman gives her a cocktail cherry on a stick. Then the barman gives a cherry to me and I say to Pepel, 'This cherry is mine', as I place it on the counter away from her. Then I tell her not to touch it unless I turn my back, in which case she can steal it, but not until she is sure that I cannot see her do so. Sometimes I have only half turned as she makes an attempt to take the cherry, when I will tell her that I can still see her. Then, when my back is completely turned she snatches the cherry and throws the stick over the counter. When I pretend to be angry while looking for my cherry, Pepel just sits there as if she knows nothing about it!

I can always tell when Pepel is happier than usual, especially when she is among people she likes and trusts. She went to everyone in turn to sit on their shoulders to 'do their hair' before we left the bar to go back to the ship. One customer said, 'I hope she will not do a Duke of Edinburgh on me'. He was referring to an incident which happened recently when a monkey urinated on the Duke as he was passing the monkey cage at a zoo.

Ray and Brian accompanied Pepel and me back to the ship, where we all sat drinking and talking far into the night. The watchman had told me that he would not be sailing after all until the next afternoon tide, so I was in no hurry to see my guests leave, although Pepel was fast asleep when the taxi arrived for them.

The next day I took Pepel to Roker, where I had promised to meet Ray and Brian. It was crowded along the seafront and it was the hottest day I had ever known in the north-east. The beach was packed with people who were taking advantage of the unusual hot spell and we could hardly move, let alone find a quiet spot to relax in. Although Pepel is used to crowds now, it is I who feel nervous when I am trapped by people who will not move as they stop to stare at Pepel.

We sat at the edge of a grass verge while Pepel enjoyed an ice-cream, until we became surrounded by people, mostly children, who had stopped to watch her making short work of her cornet. The crowd got so big that the traffic could not get by on the busy promenade. After a while we moved on to find a more secluded spot.

The rest of the day was spent by the boating lake, where we all stretched out in the long grass on the bank. There is a model train which runs around the lake, giving children a joy ride for sixpence. Pepel divided her interest between that and the boat in which Ray was sitting with his knees up, paddling it round the lake as if he was still a child. I felt sorry for Ray. He was still unemployed and, like Brian, was finding it very difficult to get a suitable job. I sat there wishing I could get jobs for them on board the *Arduity*.

As always, happy times pass quickly. The wonderful relaxation we had all enjoyed came to an end on that Easter Monday with the incoming tide. Pepel and I had to be back on board once more. Leaving the holiday crowds to their leisure, we all returned to Hendon dock, arriving there at the same time as the pilot was climbing aboard.

Ray and Brian followed the ship as it was manoeuvred through the docks and through the last bridge. Pepel sat on her rail watching them and saying cheerio in her own language. I stood beside her with a lump in my throat, wondering when we should see them again, as the *Arduity* passed through the harbour and they faded into the distance.

# Chapter Twenty-One

The Year of the Animals, an ancient Chinese system based on lunar cycles for reading horoscopes, was devised by Buddha over two thousand years ago when he summoned all the animals of the earth to his presence. When only twelve creatures turned up he rewarded them by giving each of them a 'Year' bearing its name. He also gave them power to enter into the minds of subjects born under their signs, which are the Tiger, Rat, Buffalo, Cat, Snake, Dragon, Monkey, Goat, Horse, Cock, Dog and Pig. Each animal, so the legend goes, has the power to control the destiny of its subjects.

1968 is the Year of the Monkey. Those who read their newspapers regularly may have noticed, as I have, that hardly a week has gone by without some reference has been made to incidents involving a monkey. I quote just a small selection of the articles here:

'The Cruel Fate of a Martyred Monkey', and 'Shot at

Dawn – the Monkey Who Played with Fire'. These two news items were about a poor chap called 'Bamba', a monkey living in the Yemen who accidentally knocked over a stove, setting fire to the building it was in. Poor Bamba was taken before a magistrate and sentenced to death for arson. 'Humanely, the executioner asked Bamba what his last wish was'. There was no answer. Then a sword flashed. And Bamba's severed head was left hanging on a post as a warning to fire-bugs, both human and animal. No comment!

Then there was the sad case of the poor creature who escaped from his jailers at London's Heathrow Airport as he was on his way to the vivisector. He was named 'Bimbo' by the kind airport workers who tried to help him evade recapture. They left food for him to collect when he came out of hiding at night. God bless them.

Bimbo was caught but was saved from vivisection by being sent instead to Chessington Zoo, in Surrey. He was again in the news when he escaped from the monkey compound to have a romp round the Zoo. 'Bimbo the artful dodger tries one dodge too many', said the *Daily Mirror*. The news item was illustrated by photographs of Bimbo on the outside of the compound unconcernedly pulling a face at another monkey on the inside, and from the inside after his second capture pulling a face at his spoil-sport captors. I would like to take Pepel to visit brave Bimbo one day. In the meantime, may he live a long and happy life. He deserves it.

'Don't Try to Drink a Monkey Under the Table' is the heading to an article written about 'Nancy', a ten-year-old monkey, and twenty-five male monkeys being used in experiments at the St. Elizabeth's Hospital in Washington. Scientists there are using monkeys for research into the causes and cure of alcoholism in humans. The only trouble is that so far they have been unable to make a monkey an alcoholic. They have found that although the monkey will drink and enjoy the liberal quantities of whisky they give it, unlike humans, it refuses to drink one over the eight, or have one for the road. Probably Nancy and her drinking pals realise that if they allow themselves to become drunk, their brains will be removed to see what damage has been done, for the sake of the humans who are so generous with their whisky.

Pepel has been drinking for almost five years and although the alcohol makes her sleepy, she too refuses a night cap. She will always drink greedily at first, then, realising that it is taking effect, she suddenly stops and refuses all offers to 'have just one more'.

Let us hope that Nancy and her boozing companions will continue to make monkeys out of their scientific wine and spirit merchants!

There is also the story of a family in Yorkshire who defied the high-handed local council by refusing to part with their two pet monkeys. The story appeared in the *Daily Mirror* and was headed 'A Family to Defy Ban on Pet Monkeys Pip and Charlie'. Apparently, the council did not mind dogs, cats and cage birds, but drew the line at monkeys, although Pip and Charlie were kept in a cage and were no trouble to anyone. I wrote to the family concerned to give them some encouragement in their effort to protect their pets. I told them that if they loved their pets as much as I love Pepel even the threat of eviction would seem small compared to the happiness of Pip and Charlie.

'Great Guy – the gorilla who fancied his keeper's trousers' is the title of an article written in *Reveille* about another animal lover, Laurie Smith, senior head keeper, and one of his charges at the London Zoo, Guy, the famous gorilla. Apparently, Laurie Smith was a little embarrassed one day when Guy tried to relieve him of his trousers in front of some visitors to the Zoo. Guy often played with his keeper, who spent some time catching animals in Africa and South America before he went to the London Zoo. He has a deep understanding for monkeys who, he says, 'communicate by emotions. They convey their mood in a much more reliable way than a human being, who may say one thing and mean another. I suppose you could say that animals are more honest', said Smithy, who is recognised as an authority on them in the zoological world.

The 'cupboard love' found in most other domesticated animals is never found in the monkey. Pepel has proved that to me over and over again. Her love for me is almost overpowering. She wants to be near me all the time. Sitting on my desk, she has seen practically every word written in this book, seeming to know that it is about her. No matter what our circumstances have been, Pepel has always been contented to share whatever has come our way.

The Year of the Monkey 1968 will always be a memorable one to me. God made me the happiest man alive when he restored Pepel to me after her terrible experience on board the *Arduity*. She has come a long way with me since she left her native Sierra Leone and I hope and pray that, no matter what happens in the future, Pepel and I shall remain together and share what comes our way, good or bad.

In the meantime, I dedicate this book to true animal lovers everywhere, and I hope that I shall soon be able to write *More About Pepel and People.*